RUBBER CHEMICALS

J. VAN ALPHEN

RUBBER CHEMICALS

REVISED AND ENLARGED EDITION

Edited by

C. M. VAN TURNHOUT

Plastics and Rubber Research Institute TNO, Delft, Holland

D. REIDEL PUBLISHING COMPANY

DORDRECHT-HOLLAND / BOSTON-U.S.A.

This revised and enlarged edition of J. van Alphen's **Rubber Chemicals**
is published in collaboration
with the Plastics and Rubber Research Institute TNO, Delft, Holland

The Plastics and Rubber Institute TNO is grateful
to Shell International Chemical Company Limited
for assistance in the production of this book

Library of Congress Catalog Card Number 73-81826

ISBN-13: 978-94-010-1723-7   e-ISBN-13: 978-94-010-1721-3
DOI: 10.1007/978-94-010-1721-3

Sold and distributed in the U.S.A., Canada, and Mexico
by D. Reidel Publishing Company, Inc.
306 Dartmouth Street, Boston,
Mass. 02116, U.S.A.

Sold and distributed in all other countries
by D. Reidel Publishing Company,
P.O. Box 17, Dordrecht, Holland

Preface

The ever increasing number of rubber chemicals that are being introduced has led to an increasing rate of depletion of the first edition of Dr. J. van Alphen's book which so usefully provided chemical names, trade names and the names of suppliers of a very large number of commonly used rubber additives. The need for a revised edition of this valuable book, original edited by „Rubber-Stichting" in 1956 was recognized by Shell International Chemical Company who was ready and willing to sponsor the necessary work to update and extend the contents of the book, preparations of which — following the death of Dr. van Alphen — were already taken in hand by Mr. R. J. Kuipers, at that time a TNO rubber technologist. Unfortunately Mr. Kuipers did not live to see the completion of this work and the Economic Technical Information Centre of the TNO Plastics and Rubber Research Institute was prepared to undertake the extensive task of cataloguing the expanded range of rubber chemicals in the following groups, largely following the original pattern set by Dr. van Alphen. The present book contains the following chapters:

  peptizing agents
  vulcanizing agents
  accelerators
  activators
  retarders
  blowing agents
  antidegradants
  co-agents

It is gratefully acknowledged that several producers have responded to the request for data and information, so that all chapters are now fully up to date.

The chapter on peptizers now includes newly developed products that are already active at comparatively low temperatures as well as materials displaying prolonged peptizing activity during compounding. With regard to vulcanizing agents relatively few developments have taken place and this chapter has not been changed substantially.

It goes without saying that the important groups of vulcanization accelerators and activators have been treated with considerable care.

Specific attention has been paid to include those accelerators which show different degrees of 'delayed action', so important in modern compounding, and those materials which consist of a combination of accelerators with other active ingredients resulting in special effects. The chapter on retarders reflects the comparative few developments in this field, but the number of blowing agents has considerably increased over the years, particularly because of the need for higher decomposition temperatures, especially with a view to injection moulding. Consequently this chapter has also been expanded significantly.

A new chapter has been introduced on antidegradants. They constitute a wide field of additives including antioxidants, anti-ozonants, anti-flex cracking agents, heat stabilizers and metal deactivators. This chapter has thus been significantly extended in comparison with the antioxidant section in the first edition.

In the last decade carbon black manufacturers have adopted considerable standardization with respect to types and grades of black. The corresponding rationalization has led to the envisaged simplification of the situation and carbon black manufacturers nowadays supply extensive and comprehensive data on the types now available. It has consequently been decided to omit the section on carbon black. Emulsifiers that were included in the first edition have now been left out as it was felt that ready reference should be provided to materials of general use throughout the rubber industry.

The single index on trade marks and chemical names of compounds has been replaced by two separate alphabetical indices. The original edition was in English and German, but in view of the general use of the English language in many areas of technical endeavour it was decided to issue the present edition in English only.

I trust that the book in its new form will prove to be even more useful than its predecessor, especially to technologists anywhere in the large rubber processing industry.

J. M. Goppel
Director of Research
KONINKLIJKE/SHELL PLASTICS LABORATORIUM DELFT

Abbreviations

Rubber types

Acrylic rubber	ACM
Butadiene rubber	BR
Butyl rubber	IIR
Chloroprene rubber	CR
Chlorosulfonated polyethylene	CSM
Ethylene-propylene rubber	EPM
Ethylene-propylene-diene terpolymer	EPDM, EPT
Isoprene rubber	IR
Natural rubber	NR
Nitrile-butadiene rubber	NBR
Silicone rubber	Si, PSi, CSi, PVSi
Styrene-butadiene rubber	SBR
Polyacrylate rubber	ANM
Polysulphide rubber	none

Rubber chemicals

Accelerators	Acc.
Activators	Act.
Antidegradants	Antidegr.
Antioxidants	Oxi.
Antiozonants	Ozo.
Flex-cracking agents	Flex.
Heat-stabilizers	Heat.
Metal-Poison inhibitors	Inh.
Blowing agents	Ba.
Co-agents	Ca.
Peptizers	Pept.
Retarders	Ret.
Vulcanizing agents	Vulc.

Miscellaneous

Food and Drug Administration approved	FDA appr.
Freezing Point	F.P.
Melting Point	M.P.
Melting Range	M.R.
Molecular Weight	M.W.
Specific Gravity	S.G.
No longer on selling range	†
See also	△
Delayed action	Del. act.

Names of Manufacturers

Abbreviations	Full addresses
Aceto	**Aceto Chemical Company Inc.** 126-02 Northern Blvd., Flushing, New York 11368, U.S.A.
ACNA Montecatini	**A.C.N.A. Montecatini Edison** Largo Donegani 1-2, Milano, Italy
Akron	**Akron Chemical Company** 255 Fountain Street, Akron, Ohio 44304, U.S.A.
Alkali	**Alkali Chemical Corporation India Ltd.** Rishra, West Bengal, India
Allied	**Allied Chemical Corporation, Plastics Div.** P.O. Box 365, Morristown, New Jersey 07960, U.S.A.
Cyanamid	**American Cyanamid Company, Rubber Chemicals Dept.** Bound Brook, New Jersey 08805, U.S.A.
American Hoechst	**American Hoechst Corp.** 4331 Chesa Peake Drive, Charlotte, N.C. 28208, U.S.A.
Anchor	**Anchor Chemical Co. Ltd.** Clayton, Manchester M 11 4 SR, England
Arwal	**Arwal Chemicals Inc.** 1961 West Market Street, P.O. Box 429, Akron, Ohio 44309, U.S.A.
Ashland	**Ashland Chemical Comp.** P.O. Box 1503, Houston, Texas 77001, U.S.A.
BASF	**Badische Anilin & Soda Fabrik A.G.** 6700 Ludwigshafen/Rhein, Germany
Bayer	**Bayer A.G.** 509 Leverkusen, Germany
Bennett	**D. G. Bennett Chemicals** 11A St. John's Hill, London SW 11 . 1TS, England
Bozzetto	**Bozzetto Industrie Chimiche S.p.A.** P.O. Box 240, Pedrengo (Bergamo), Italy

Abbreviations	Full addresses
CdF	**CdF Chimie** Cedex No. 5, Tour Aurore, 92 Paris, France
Chemko	**Chemko** Strázske, Czechoslovakia
Ciba Geigy	**Ciba Geigy A.G.** CH-4000 Basel 7 (Rosenthal), Switzerland
Conestoga	**Conestoga Chem. Corp., Div. of Chemetron** Wilmington Ind. Park, Foot of East 7th Str. Wilmington, Delaware 19801, U.S.A.
Degussa	**Degussa** P.O. Box 3993, 6000 Frankfurt/Main 1, Germany
Dimitrova	**Chemické závody Juraja Dimitrova** Bratislave, Czechoslovakia
Du Pont	**E.I. Du Pont de Nemours & Co. Inc.** Wilmington, Delaware 19898, U.S.A.
Durham	**The Durham Chemical Group** Birtley, County Durham, England
Eastman	**Eastman Kodak Company,** **Tennessee Eastman Comp. Div.** Kingsport, Tennessee 37662, U.S.A.
Fairmount	**Fairmount Chemical Co. Inc.** 117 Blanchard Street, Newark, New Jersey 07105, U.S.A.
Fisons	**Fisons Industrial Chemicals Ltd.** Willows Works, Derby Road, Loughborough, Leicestershire, England
Goodrich	**B. F. Goodrich Chem. Comp.** 3135 Euclid Avenue, Cleveland, Ohio 44115, U.S.A.
Goodyear	**The Goodyear Tire and Rubber Company** Akron, Ohio 44316, U.S.A.
Grandel	**Deutsche Oelfabrik Dr. Grandel & Co.** P.O. Box 111929, 2 Hamburg 11, Germany
Hall	**The C. P. Hall Company** 4460 Hudson Drive, Stow, Ohio 44224, U.S.A.

Abbreviations	Full addresses
Hasselt	**Van Hasselt N.V.** Amsterdamseweg 18, Amersfoort, The Netherlands
Hercules	**Hercules Inc. Intern. Dept.** Wilmington, Delaware 19899, U.S.A.
ICI America	**ICI America Inc.** Stamford, Connecticut, U.S.A.
Icianz	**Icianz Ltd.** Melbourne 3001, Australia
ICI (India)	**ICI (India) Pty. Ltd.** Chowringhee, Calcutta, India
ICI	**Imperial Chemical Industries Ltd.** Millbank, London SWIP 3JF, England
Ticino	**Industria Chimica del Ticino S.p.A.** Via del Porto, 28040 Marano Ticino, Italy
Kenrich	**Kenrich Petrochemicals Inc.** Foot of East 22nd Street, Bayonne, New Jersey 07002, U.S.A.
Lucidol	**Lucidol, Div. of Wallace & Tiernan Inc.** 1740 Military Road, Buffalo, New York 14240, U.S.A.
May & Baker	**May & Baker Ltd.** Dagenham, Essex RM 10 7 XS, England
Merck	**Merck & Co. Inc.** Rahway, New Jersey 07651, U.S.A.
Metallgesellschaft	**Metallgesellschaft A.G., Abt. TA** Reuterweg 14, 6000 Frankfurt/Main, Germany
Monsanto	**Monsanto Corp.** 800 N. Lindbergh Blvd. St. Louis, Missouri 63166, U.S.A.
Nat. Polychemicals	**National Polychemicals Inc.** Wilmington, Massachusetts 01887, U.S.A.
Naugatuck S.p.A.	**Naugatuck S.p.A.** Corso Vinzaglio 35, 10121 Torino, Italy
Neville	**Neville Chemical Comp.** Neville Island, Pittsburgh, Pa. 15225, U.S.A.

Abbreviations	Full addresses
Neville Synthèse	**Neville Synthèse Organics Inc.** Neville Island, Pittsburgh, Pa. 15225, U.S.A.
Newalls	**Newalls Insulation and Chemical Co. Ltd.** Washington, Co. Durham, England
Norac	**The Norac Co. Inc.** 405 South Motor Avenue, P.O. Box F, Azusa, California 91703, U.S.A.
Nourychem	**Nourychem Corp.** 2153 Lockport-Olcott Road, Burt, New York 14028, U.S.A.
Noury	**Noury & v. d. Lande N.V. *** P.O. Box 10, Deventer, The Netherlands
Ouchi Shinko	**Ouchi Shinko Chemical Industrial Co. Ltd.** 3-7, 1-chome, Kobuna-cho, Nihombashi, Chuo-ku, Tokyo, Japan
Pearson	**William Pearson Ltd.** Clough Road, Bull, Yorkshire, England
Pennwalt	**Pennwalt Corp.** 900 First Avenue, King of Prussia, Pa. 19406, U.S.A.
Pitt Consol	**Pitt Consol Chemical Company** 380 North Street, Teterboro, New Jersey 07608, U.S.A.
Polychimie	**Polychimie** Le Pressoir Vert, 45 Semoy, (Orléans), France
Prochim	**Prochim** 59 Courchelettes, P.O. Box 236, Douai, France
Raschig	**Dr. F. Raschig GmbH** 6700 Ludwigshafen/Rhein, Germany
Reagens	**Reagens, Societá per Azioni, Industria Chimica** S. Giorgio di Piano, Bologna, Italy
Reichhold Albert	**Reichhold Albert Chemie A.G.** Harvestehuderweg 18, 2 Hamburg 13, Germany
* Change of name:	**AKZO CHEMIE B.V.** Sales office Deventer

Abbreviations	Full addresses
Reichhold	**Reichhold Chemicals Inc.** RCI Building, White Plains, New York 10602, U.S.A.
Rhein-Chemie	**Rhein-Chemie Rheinau GmbH** P.O. Box 84 and 104, 6800 Mannheim 81, Germany
Rhône Poulenc	**Société des Usines Chimiques Rhône Poulenc** 22, Avenue Montaigne, 75 Paris VIIIe, France
Robinson	**Robinson Brothers Ltd.** Ryders Green, West Bromwich, Staffordshire, England
Ross	**Frank B. Ross Co. Inc.** 6-10 Ash Street, Jersey City, New Jersey 07304, U.S.A.
Rubber Regenerating	**The Rubber Regenerating Comp. Ltd.** Trafford Park, Manchester M 17 1 DT, England
Sanshin	**Sanshin Chemical Industry Co. Ltd.** Yanai Minato, Yanai City, Yamaguchi Pref., Japan
Sartomer	**Sartomer Resins Inc.** P.O. Box 56, Essington, Pa. 19029, U.S.A.
Schill & Seilacher	**Schill & Seilacher Chem. Fabrik** Moorfleeterstrasse 28, P.O. Box 460, 2000 Hamburg 74, Germany
Schlickum	**Schlickum Werke J. Schlickum & Co.** P.O. Box 225, 2000 Hamburg 36, Germany
Shell	**Shell International Chemical Company Ltd.** Shell Centre, London S.E.1. 7PG, England
Organo Synthèse	**Société Française d'Organo Synthèse S.A.** 159, Avenue de Roule, 92 Neuilly/Seine, France
Spolek	**Spolek Pro Chemikou A Hutni Vyrobu** Ustí nad Labem, Czechoslovakia
Stauffer	**Stauffer Chemical Company** 380 Madison Avenue, New York, New York 10017, U.S.A.
Sumitomo	**Sumitomo Chemical Co. Ltd.** 15, 5-chome, Kitahama, Higashi-ku, Osaka, Japan

Abbreviations	Full addresses
Uniroyal	**Uniroyal Inc.** Spencer Street, Naugatuck, Connecticut 06770, U.S.A.
UOP	**Universal Oil Products Company** State Highway 17, East Rutherford, New Jersey 07073, U.S.A.
Vanderbilt	**R. T. Vanderbilt Company Inc.** 230 Park Avenue, New York, New York 10017, U.S.A.
Vondelingenplaat	**Fabriek van Chemische Producten** **Vondelingenplaat N.V.** P.O. Box 7120, Rotterdam, The Netherlands
Vychodoceské	**Vychodoceské Chemické závody** Kolin, Czechoslovakia
Wallace & Tiernan	**Wallace & Tiernan Chemie GmbH** P.O. Box 149, 887 Günzburg/Donau, Germany

Accelerators

INORGANIC

1 1. Lead oxides

2. —

3.

Mix Lpb 80	Bozzetto a
Mix Pb 80	Bozzetto b
Polyminium	Polychimie b
Polytharge	Polychimie a
grades A, B, D	
RC Granulat PbO	Rhein Chemie a
RC Granulat Pb_3O_4	Rhein Chemie b
—	Anchor c

4. a. PbO dispersion. Acc. for CR; act. for NR, NBR, SBR, IIR; vulc. for CR and CSM.
    b. Pb_3O_4 dispersion. Acc. for CR; act. for IIR,
    c. PbO powder (litharge). Vulc. for CR.
    △ Act. 1
    △ Vulc. 1

ORGANIC

Aldehyde-ammonia compounds

2 1. Hexamethylenetetramine (with additives)

2. $(CH_2)_6N_4$

3.

Aceto HMT	Aceto
Eveite UR	ACNA Montecatini
Herax UTS	Dimitrova
Hexa K	Degussa
Hexalit	Chemko
grades S,01,02	
Nocceler H	Ouchi Shinko
RC Granulat Hexa	Rhein Chemie
(80% hexa and 20% saturated hydrocarbons and dispersion agents)	

Rhenocure Hexa	Rhein Chemie †
Sanceler H	Sanshin
Soxinol H	Sumitomo
Vulkacit H 30	Bayer

4. Hygroscopic white powder. M.W. 140.19. Sublimates at 263 °C. S.G. 1.3.

Acc. for NR, SBR, NBR; act. for mercapto-, sulphenamide-, thiuram- and zinc dithiocarbamate-type acc.

△ Act. 15

3 1. Ethylchloride - formaldehyde - ammonia reaction product
   2. —
   3.

Trimene Base	Uniroyal
Trimene	Uniroyal a
(Trimene Base with stearic acid 1 : 1)	
Vulcafor EFA	ICI

   4. Dark brown viscous liquid. S.G. 1.09.
      a. Thick paste, S.G. 0.97.
      Acc. and latex foam stiffener for NR, SBR.

Aliphatic amines and their condensation products

4 1. Aldehyde-amine condensation product
   2. —
   3. Vulcafor PT   ICI †
   4. Brown resinous mass. S.G. 1.05.

5 1. Alkylamine
   2. —
   3. Rhenocure PA   Rhein Chemie
   4. Light coloured viscous liquid. B.R. 215–225 °C.; S.G. 0.93.
      Semi-ultra acc. Crosslinking agent for acrylate elastomers.

6 1. Butyraldehyde - monobutylamine condensation product
   2. —
   3. Accelerator 833   Du Pont

4. Reddish amber liquid. F.P. 116 °C.; S.G. 0.86.
   Staining acc. for NR, SBR and reclaim. Also for latex. Acc. for self-curing CR cements.

7   1. Cyclohexyl ethyl amine
   2. $(CH_2)_5CH.NH.C_2H_5$
   3. Vulkacit HX   Bayer
   4. Pale yellow liquid. M.W. 127; B.P. 165 °C.; S.G. 0.85.
      Sec. acc. and act.
      △ Act. 12

8   1. Dibutylamine
   2. $(C_4H_9)_2NH$
   3. —   B.A.S.F.
   4. Clear liquid. S.G. 0.75.
      Sec. acc. for Vulkacit P extra N (Bayer).

9   1. Polyethylene-polyamine
   2. —
   3. Vulkacit TR   Bayer
   4. Yellow to reddish-brown liquid. S.G. 0.99.
      Semi-ultra acc. used alone or in combination with Vulkacit 1000.

10   1. Tricrotonylidenetetramine (with dispersion agents)
   2. —
   3. Vulkacit CT-N   Bayer
   4. Viscous brown oil. S.G. 1.05.
      Acc. for NR, IR, BR, SBR, NBR.

Aromatic amines and their condensation products

---

11   1. Acetaldehyde-aniline condensation products
   2. —
   3. Crylene   Uniroyal a
      (mixture with stearic
      acid 67 : 33)
      Nocceler K   Ouchi Shinko b
      VGB   Uniroyal c
      Vulcafor RN   ICI †

4. a. Thick brown paste. S.G. 1.014.
   b. Reddish-brown powder. M.P. 55 °C.
   c. Brown resinous powder. M.R. 60–80 °C.; S.G. 1.15.
   Staining acc. for NR; oxi, heat for NR.
   △ Antidegr. 40

12 1. Anhydroformaldehyde - aniline condensation product
   2. —
   3. Vulcafor MA   ICI †
   Vulcafor MA   ICI (India) †
   4. Yellow powder. M.P. 133 °C.

13 1. Anhydroformaldehyde p-toluidine condensation product
   2. —
   3. Vulcafor MT   ICI †
   4. White powder. M.P. 133 °C.

14 1. Butyraldehyde - aniline condensation product
   2. —
   3. Accelerator 21   Anchor
   Antox Special   Du Pont
   Beutene   Uniroyal
   Butanyl-1   Ticino
   Nocceler 8   Ouchi Shinko
   Rapid Accelerator 300A   Rhône Poulenc
   Vulcafor BA   ICI
   4. Orange-red oily liquid. S.G. 0.94–1.02.
   Semi-ultra acc. for Neoprene W and hydrocarbon rubber; act.
      for thiazoles, thiurams and guanidines; antidegr. for CR.
   △ Act. 11
   △ Antidegr. 41

15 1. 4,4'-Diamino diphenyl methane
   2. $CH_2(C_6H_4.NH_2)_2$
   3. Robac 4.4   Robinson a
   Tonox   Uniroyal b
   4. a. Light brown powder. M.W. 198.3; M.R. 75–85 °C.
         Acc. for CR; ret. for IIR; anti-frosting agent for NR.
      b. Brown waxy lump. S.G. 1.18.
   △ Antidegr. 42
   △ Ret. 2

16 1. N,N-Di-(N'-ethylidene-anilino) aminobenzene
2.   C₆H₅
                         |
      C₆H₅–NH–HC–N–CH–NH–C₆H₅
                 | |
                CH₃ CH₃

$$C_6H_5-NH-HC-\underset{\underset{CH_3}{|}}{N}-\underset{\underset{CH_3}{|}}{CH}-NH-C_6H_5$$

3. Eveite A   ACNA Montecatini
4. Brown oily liquid. M.W. 331.45; S.G. 1.06–1.07.
   Acc. and antidegr.
   △ Antidegr. 43

17 1. Heptaldehyde - aniline condensation product
2. —
3. Heptene Base   Uniroyal
4. Dark brown liquid. S.G. 0.92.
   Acc. for NR; act. for thiazole- and thiuram-type acc.
   △ Act. 14

18 1. Butyraldehyde - butylidene-aniline reaction product
2. —
3. A-32   Monsanto a
   Eveite 101   ACNA Montecatini b
4. a. Dark-amber liquid. S.G. 1.01.
      Used alone or in combination with thiazoles or thiurams.
          FDA appr.
   b. Brown liquid. M.W. 201.30; S.G. 0.95.
      Staining.

19 1. Butyraldehyde - acetaldehyde - aniline reaction product
2. —
3. A-100   Monsanto
4. Dark red-brown, oily liquid. S.G. 1.01–1.07.
   Acc. for NR, hard rubber, synth. rubbers.

20 1. Homologous acroleines - aromatic bases condensation products
2. —
3. Vulkacit 576   Bayer
4. Reddish-brown liquid. S.G. 0.99.
   Staining acc. for NR, IR, BR, SBR, NBR.

Thioureas

21   1. Mixed di-alkyl thiourea
2. —
3. Pennzone L   Pennwalt
   Pennzone L   Vondelingenplaat
4. Amber-coloured liquid. S.G. 1.00.
   Acc. for CR, epichlorohydrinrubber, and for mixtures of CR and chlorobutylrubber.

22   1. N,N'-Dibutylthiourea
2. $(C_4H_9NH)_2CS$
3. Accelerator DBT   BASF
   DBTU   Prochim
   —   Degussa
   Pennzone B   Pennwalt
   Pennzone B   Vondelingenplaat
   Robac DBTU   Robinson
4. Off-white powder. M.W. 188.3; M.P. 65 °C.
   Acc. for mercaptan-modified CR; act. for EPDM and NR; antidegr. for NR-latex and thermoplastic SBR.
   △ Act. 35
   △ Antidegr. 109

23   1. 1,3-Diethylthiourea
2. $(C_2H_5NH)_2CS$
3. DETU   Prochim
   JOR 4050   Bozzetto
   —   Degussa
   Pennzone E   Pennwalt a
   Pennzone E   Vondelingenplaat a
   RC Granulat DETU   Rhein Chemie
     (80 % DETU, 20 % saturated hydrocarbons and special dispersion agents)
   Robac DETU   Robinson

4. Yellow powder. M.W. 132.2; M.P. 75 °C.; S.G. 0.98.
   a. White flakes. S.G. 1.12.
   Acc. for mercaptan-modified CR (Neoprene W); antidegr. for
     NR, NBR, SBR, CR.
   Water soluble.
   △ Antidegr. 110

24 1. Dimethyl-ethyl-thiourea
    2. $C_2H_5NHCSN(CH_3)_2$
    3. Thiate B   Vanderbilt
    4. Reddish-brown liquid. M.W. 132; S.G. 1.05.
      Acc. for Neoprene-W compounds.

25 1. Sym. Diphenyl-thiourea
      syn. Thiocarbanilide
    2. $C_6H_5.NH.CS.NH.C_6H_5$
    3. A–L Thiocarbanilide   Monsanto
      DPTU   Prochim
      Eveite TC   ACNA Montecatini
      Nocceler C   Ouchi Shinko
      Soxinol C   Sumitomo
      Stabilisator C   Bayer
        (formerly Vulkacit CA)
      —   Degussa
      Vulcafor TC   ICI
    4. Cream-white powder. M.W. 228; M.P. 149 °C. min.; S.G. 1.31.
      Non-staining sec. acc. for CR, EPDM.
      △ Act. 37

26 1. N,N'-Di-ortho-tolyl-thiourea
    2.

3. Eveite DOT   ACNA Montecatini
   Nocceler DOTU   Ouchi Shinko
4. Grayish-white powder. M.W. 256.36; M.P. 149 °C.
   Acc. for NR, CR.

27 1. Ethylene-thiourea
   syn. 2-Mercaptoimidazoline
2. H_2C—N
   H_2C C-SH
   N
   H

3. Accelerator MI 12   Metallgesellschaft
   —   Degussa
   E.T.U.   Hasselt
   E.T.U.   Prochim
   JOR 4022   Bozzetto
   JOR 4022 oleato   Bozzetto
       (83 % JOR 4022,
       17 % paraffin oil)
   Na-22   Du Pont
   Na-22D   Du Pont
       (80 % dispersion of
       NA-22 in oil)
   Nocceler 22   Ouchi Shinko
   Pennac CRA   Pennwalt
   Pennac CRA   Vondelingenplaat
   Robac 2.2   Robinson
   Rodanin S-62   Dimitrova
   Sanceler 22   Sanshin
   Soxinol 22   Sumitomo
   Vulkacit NPV/C   Bayer
       (coated)
4. White powder. M.W. 102.16; M.P. 195 °C. min.; S.G. 1.43–1.45.
   Non-staining acc. for CR; ozo for NR.
   △ Antidegr. 93

28   1. Tetramethylthiourea

     2.
$$\begin{array}{c} N(CH_3)_2 \\ \diagup \\ C = S \\ \diagdown \\ N(CH_3)_2 \end{array}$$

     3. NA-101   Du Pont

     4. Light tan flakes. M.W. 132; M.P. 75–80 °C.; S.G. 1.2. Non-staining acc. for CR.

29   1. Trimethylthiourea

     2. $CH_3NHCSN(CH_3)_2$

     3. Nocceler TMU   Ouchi Shinko

        Thiate E   Vanderbilt

     4. Light tan flakes. M.W. 119; M.R. 68–78 °C.; S.G. 1.23. Acc. for Neoprene W-compounds.

Guanidine-derivatives

30   1. Diarylguanidine blend

     2. R1.NH.C(:NH).NH.R2

     3. Accelerator 49   Cyanamid

     4. White to pinkish-white powder. M.P. ca. 134 °C.; S.G. 1.20. Acc. for NR; act. for MBT or MBTS in SBR.

       △ Act. 33

31   1. N,N'-Diphenylguanidine

     2. $HN = C(NH.C_6H_5)_2$

     3.

Accicure DPG	Alkali
Denax	Dimitrova
Denax DPG	Vychodočeské
DPG	Cyanamid
DPG	Anchor
DPG	Monsanto
DPG	Rhône Poulenc
Eveite D	ACNA Montecatini
Nocceler D	Ouchi Shinko
Pennac DPG	Pennwalt
Sanceler D	Sanshin
Soxinol D	Sumitomo

Vulcafor DPG (surface treated)	ICI
Vulcafor DPG	ICI (India)
Vulkacit D	Bayer

4. White crystalline non-hygroscopic powder or paste. M.W. 211.26; M.R. 144–146 °C.; S.G. 1.19.
Medium acc. for use with thiazoles and sulphenamides.
△ Act. 36

32 1. N,N'-Di-ortho-tolylguanidine
2. $2\text{-}CH_3.C_6H_4.NH.C(:NH)NH.C_6H_4.CH_3\text{-}2$
3.

DOTG	Anchor
DOTG	Cyanamid
DOTG	Du Pont
DOTG	Rhône Poulenc
Eveite DOTG	ACNA Montecatini
Nocceler DT	Ouchi Shinko
Soxinol DT	Sumitomo
Vulcafor DOTG	ICI
Vulcafor DOTG (surface coated)	ICI
Vulkacit DOTG	Bayer
Vulkacit DOTG/C (surface coated)	Bayer

4. White powder. M.W. 239; M.R. 167–173 °C.; S.G. 1.19.
Slow curing acc. for NR, SBR, NBR; act. for acidic and neutral acc.
△ Act. 38

33 1. Di-ortho-tolylguanidine salt of dicatecholborate
2.

3. Nocceler PR   Ouchi Shinko
   Permalux   Du Pont
4. Grayish-brown powder. M.W. 482.8; M.P. 165 °C.; S.G. 1.25.
   Non-staining acc. for Neoprene G-types; antidegr. for NR, SBR.
   △ Antidegr. 116

34 1. ortho-Tolylbiguanide
2. $H_2N-C(:NH)-NH-C(:NH)-NH-C_6H_4.CH_3$
3. Accelerator 80   Du Pont
   Eveite 1000   ACNA Montecatini
   Nocceler BG   Ouchi Shinko
   Vulkacit 1000   Bayer
   Vulkacit 1000/C   Bayer
     (surface coated)
4. White powder. M.W. 191.24; M.P. 140 °C.; S.G. 1.2.
   Acc. for NR, IR, BR, SBR, NBR; act. for zinc dithiocarbamates,
   thiuram-, mercapto-, and sulphenamide-type acc.
   △ Act. 40

35 1. N,N',N''-Triphenylguanidine
2.
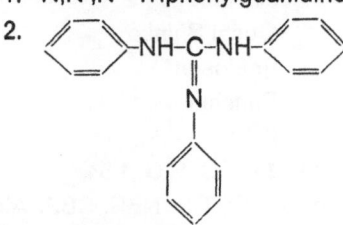

3. Vulcafor TPG   ICI †
4. White amorphous powder. M.W. 279; M.P. 142 °C.; S.G. 1.17.

Xanthates

36 1. Sodium isopropylxanthate
2. $\begin{matrix} H_3C \\ >CHO-C-S-Na \\ H_3C \| \\ S \end{matrix}$
3. Aero Xanthate 343   Cyanamid a
   Sanceler SX   Sanshin b
   Vulcafor SPX   ICI †

4. a. Light orange powder. M.W. 158; M.P. 126 °C.; S.G. 1.40.
   b. Hydrate (2 H$_2$O). M.W. 194.25; M.P. 124 °C.
   Water-soluble ultra-acc. for NR, NBR, SBR, CR. Also for latex.

37
1. Zinc dibutylxanthate
2. (H$_9$C$_4$.O.CS.S–)$_2$Zn
3. Nocceler ZBX   Ouchi Shinko
   Sanceler ZBX   Sanshin
4. White powder. M.W. 363.88; M.P. 105 °C. min.; S.G. 1.24–1.40.
   Ultra-acc. for NR, NBR, SBR. Also for latex. Non-staining.

38
1. Zinc diethylxanthate
2. (CH$_3$.CH$_2$.O.CS.S–)$_2$Zn
3. Eveite XZ   ACNA Montecatini
4. White powder. M.W. 307.

39
1. Zinc diisopropylxanthate
2. [(CH$_3$)$_2$CH.O.CS.S–]$_2$Zn
3. Accelerator ZIX   Anchor
   Nocceler ZIX   Ouchi Shinko
   Robac ZIX   Robinson
   Sanceler ZX   Sanshin
   Vulcafor ZIX   ICl †
4. White powder. M.W. 335.83. M.P. 145 °C. S.G. 1.54.
   Fast acc. at room-temperature for NR, CR, NBR, SBR. Also for
   latex.

Dithiocarbamates

40
1. Activated dithiocarbamates
2. —
3. Ancazate SM   Anchor
   Ancazate XX   Anchor
   Butyl Eight   Vanderbilt
   Merac   Pennwalt
   Merac   Vondelingenplaat
   Robac Gamma   Robinson

Royalac 133	Uniroyal
Setsit 5	Vanderbilt
Setsit 9	Vanderbilt
Setsit 51	Vanderbilt
Trimate L (50 % active)	Vanderbilt

4. Amber-to-brown liquids. S.G. ca. 1.00.
   Room-temp. acc. spec. for CR latices.

41
1. Bismuth dimethyldithiocarbamate
2. [(CH₃)₂N.CS.S–]₃Bi
3.

BDMC	Hasselt
Bismate	Vanderbilt
JO 6000	Bozzetto
Robac Bi.D.D.	Robinson

4. Yellow powder and pellets. M.W. 569.66; M.P. above 227 °C.
   (with decomposition); S.G. 2.02.
   Acc. for NR, SBR, IIR; act. for thiazole- and sulphenamide-type
   acc.
   △ Act. 18

42
1. Cadmium diethyldithiocarbamate
2. [(C₂H₅)₂N.CS.S–]₂Cd
3.

Cadmate	Vanderbilt a
Robac C.D.C.	Robinson b

4. Cream coloured powder. M.W. 409; S.G. 1.39
   a. M.R. 68–76 °C.
   b. M.P. 242 °C.
   Ultra-acc. for IRR. Acc. for EPDM.

43
1. Cadmium pentamethylenedithiocarbamate
2. [(CH₂)₅N.CS.S–]₂Cd
3. Robac C.P.D.   Robinson
4. Pale yellow powder. M.W. 433; M.P. 260 °C.
   Del.act. acc. for NR, SBR. (cements and proofing compounds).

44
1. Copper dimethyldithiocarbamate
2. [(CH₃)₂N.CS.S–]₂Cu

3. CDMA   Hasselt
   Cumate   Vanderbilt
   Hermat Cu   Dimitrova
   JO 4015   Bozzetto
   Nocceler TTCU   Ouchi Shinko
   Robac Cu.D.D.   Robinson
   Sanceler TTCU   Sanshin
   Soxinol MK   Sumitomo

4. Dark brown powder. M.W. 303.98; M.P. above 325 °C. (with decomposition); S.G. 1.75.
   Acc. for SBR, IIR; act. for thiazole- and sulphenamide-type acc.
   △ Act. 19

45 1. 2-Benzothiazyl-N,N-diethyldithiocarbamate
       syn. 2-Benzothiazyl-N,N-diethylthiocarbamylsulphide

2.

3. Ethylac   Pennwalt
   Ethylac   Vondelingenplaat
   Nocceler 64   Ouchi Shinko

4. Yellow powder. M.W. 282.45; M.P. ca. 70 °C.; S.G. 1.27.
   Non-staining acc. and act. for NR, BR, SBR, NBR, IR.
   △ Act. 17

46 1. Dibutylammonium-dibutyldithiocarbamate
       2. $(C_4H_9)_2.N.CS.S.NH_2(C_4H_9)_2$
       3. Robac D.B.U.D.   Robinson
       4. Brown flakes. M.W. 334.6; M.P. 45 °C.
          Acc. for NR and SBR. (proofing compounds); oxi for rubber-based adhesives. Non-staining.
          △ Antidegr. 100

47 1. Diethylammonium-diethyldithiocarbamate
       2. $(C_2H_5)_2N.CS.S.NH_2(C_2H_5)_2$
       3. Ancazate WSE   Anchor a
          Vulcafor DDCN   ICI †

4. Cream crystalline powder. M.W. 208; M.P. 80 °C.; S.G. 1.11.
   a. 30 % aquous solution. Yellow liquid. S.G. 1.028.
   Fast acc. for latex compounds.

48 1. Dimethylammonium-dimethyldithiocarbamate
   2. $[(CH_3)_2N.CS.S-]NH_2(CH_3)_2$
   3. Pennac DDD   Pennwalt
      Vondac DADD   Vondelingenplaat
   4. 40 % aquous solution. Pale yellow liquid. S.G. 1.055.
      Acc. for latex.

49 1. N,N-Dimethylcyclohexamine salt of dibutyldithiocarbamic acid
   2. —
   3. RZ-100   Monsanto
   4. Yellow-orange lumps. Crystallization Point 52.5 °C. min.
      Non-staining acc. for curing at room temperature of NR- and
      SBR cements and latices.

50 1. 2,4-Dinitrophenyl-dimethyldithiocarbamate
   2. $(CH_3)_2N.CS.S.C_6H_3(NO_2)_2-2,4$
   3. Safex   Uniroyal
   4. Yellow crystalline powder. M.W. 287; M.R. 140–145 °C.; S.G.
      1.57.
      Del.act. acc.; act. for thiazole-type acc.
      △ Act. 20

51 1. Ferric dimethyldithiocarbamate
   2. $[(CH_3)_2N.CS.S-]_3Fe$
   3. Nocceler TTFE   Ouchi Shinko
      Sanceler TTFE   Sanshin
   4. Black-brown powder, M.W. 416.50; M.P. 230 °C. min.
      Ultra-acc. for NBR, SBR.

52 1. Lead dimethyldithiocarbamate
   2. $[(CH_3)_2N.CS.S-]_2Pb$
   3. JO 4014   Bozzetto
      LDMC   Hasselt
      Ledate   Vanderbilt
      Robac L.M.D.   Robinson

4. White powder. M.W. 447.65; M.P. above 310 °C.; S.G. 2.43.
   Acc. for NR, BR, IR, SBR, IIR; act. for thiazole- and sulphen-
      amide-type acc.
   △ Act. 21

53 1. Lead pentamethylenedithiocarbamate
2. $[(CH_2)_5N.CS.S-]_2Pb$
3. Robac L.P.D.   Robinson
4. Off-white powder. M.W. 527.8; M.P. 230 °C.
   Del.act. acc. for NR, SBR. (continuous vulcanization of cable-
      compounds).

54 1. Piperidinium-pentamethylenedithiocarbamate
   syn. N-Pentamethyleneammonium-N-pentamethylenedithio-
      carbamate
2. $(CH_2)_5N.CS.S.H_2N(CH_2)_5$
3. Accelerator 552   Du Pont
   Accelerator 2P   Anchor
   Nocceler PPD   Ouchi Shinko
   Pentalidine   Prochim
   Robac P.P.D.   Robinson
   Vulkacit P   Bayer
4. Cream powder. M.W. 246; M.P. 175 °C.; S.G. 1.19.
   Acc. for NR, NBR, SBR; act. for thiuram- and thiazole-type acc.;
      pept. for Neoprene G and KNR types.
   △ Act. 22
   △ Pept. 6

55 1. Pipecolin methyl-pentamethylenedithiocarbamate
2.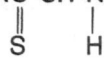
3. Nocceler P   Ouchi Shinko
4. M.W. 274.49; M.P. 120 °C.
   Acc. for NR, IR, BR, SBR, NBR, IIR.

56 1. Potassium di-N-butyldithiocarbamate
2. $(C_4H_9)_2N.CS.S.K$
3. Butyl Kamate   Vanderbilt
4. Straw coloured aquous-solution. S.G. 1.10.
   Acc. for latex.

57 1. Selenium diethyldithiocarbamate
        2. $[(C_2H_5)_2N.CS.S-]_4Se$
        3. Ethyl Selenac   Vanderbilt
           Ethyl Seleram SA-66-1   Pennwalt
             (oiled)
           Seleniame   Prochim
           Soxinol SE   Sumitomo
        4. Yellow-orange powder. M.W. 672; M.R. 59–85 °C.; S.G. 1.32.
           Acc. for IIR; vulc. for NR, NBR, SBR; act. for thiazole-type acc.
           △ Act. 23
           △ Vulc. 40

58 1. Selenium dimethyldithiocarbamate
        2.

$$\left[\begin{array}{c} CH_3 \\ {>} N{-}C{-}S{-} \\ CH_3 \quad \underset{S}{\overset{\|}{}} \end{array} \right]_4 Se$$

        3. Methyl Selenac   Vanderbilt
        4. Yellow powder and rods. M.W. 559.78; M.R. 140–172 °C.; S.G.
             1.58.
           Acc. and vulc. for NR, IIR, SBR, BR, IR.
           △ Vulc. 41

59 1. Sodium dibutyldithiocarbamate
        2. $(C_4H_9)_2N.CS.S.Na$
        3. Ancazate WSB   Anchor
             (48 % aquous sol.)
           Butyl Soderame   Prochim
             (47 % aquous sol.)
           Nocceler TP   Ouchi Shinko
           Robac S.B.U.D.   Robinson
             (45 % aquous sol.)
           SBTC   Bozzetto
             (40 % aquous sol.)
           Soxinol TP   Sumitomo
             (40 % aquous sol.)
           Tepidone   Du Pont
             (47 % aquous sol.)

4. Clear brown liquid. M.W. 227; S.G. 1.075–1.09.
   Ultra-acc. for latices and for reclaimed mixes; act. for thiazole-
   type acc.
   △ Act. 24

60 1. Sodium diethyldithiocarbamate
2. $(C_2H_5)_2N.CS.S.Na$
3.

Ethyl Soderame	Prochim a
Eveite L	ACNA Montecatini b
Nocceler SDC	Ouchi Shinko a
Pennac SDED	Pennwalt
(25 % aquous sol.)	
Robac S.E.D.	Robinson
(23 % aquous sol.)	
Sanceler ES	Sanshin
(20–22 % aquous sol.)	
Soxinol ESL	Sumitomo
(18–22 % aquous sol.)	
Super Accelerator 1500	Rhône Poulenc c
Vondac SDED	Vondelingenplaat
(25 % aquous sol.)	
Vulcafor SDC	ICI †

4. a. White crystalline powder. M.W. 171.26; S.G. 1.3.
   b. White crystalline powder. M.W. 207.29; M.P. 90–92 °C.; S.G.
      1.30.
   c. Hydrate. M.W. 225; S.G. 1.30.
   Ultra-acc. for NR- and SBR latices; act. for guanidine-type acc.
   Non-staining.
   △ Act. 25

61 1. Sodium dimethyldithiocarbamate
2. $(CH_3)_2N.CS.S.Na$
3.

Eveite K	ACNA Montecatini
Methyl Soderame	Prochim
Nocceler S	Ouchi Shinko
Sanceler S	Sanshin
Soxinol MSL	Sumitomo

4. 40–42 % aquous-solution, light orange coloured. M.W. 143.20; S.G. 1.17–1.19.
   Ultra-acc. for NR- and SBR latices in combination with other acc.

62 1. Tellurium diethyldithiocarbamate
2. [(C₂H₅)₂N.CS.S–]₄Te

3. Soxinol TE	Sumitomo
TDEC	Hasselt
Tellurac	Vanderbilt
Tellurame	Prochim

4. Orange-yellow powder and rods. M.W. 720.69; M.R. 108–118 °C.; S.G. 1.44.
   Acc. for NR, SBR, NBR, IIR, EPDM; act. for thiazole- and thiuram-type acc.
   △ Act. 26

63 1. Zinc dithiocarbamate (activated)
2. —
3. Ancazate Q   Anchor
4. Viscous brown liquid. S.G. 1.10.
   Room-temperature acc. for NR, SBR.

64 1. Zinc dibenzyldithiocarbamate
2. [(C₆H₅CH₂)₂N.CS.S–]₂Zn

3. Arazate	Uniroyal a
Robac Z.B.E.D.	Robinson b
ZBEC	Hasselt b

4. Creamy-white powder. M.W. 610.2; S.G. 1.41.
   a. M.R. 160–175 °C.
   b. M.P. 182 °C. min.
   Ultra-acc. for IIR, SBR, NR. Also for latex. Act. for thiazole- and sulphenamide-type acc.
   △ Act. 27

65 1. Zinc dibutyldithiocarbamate
2. [(C₄H₉)₂N.CS.S–]₂Zn

3. Aceto ZDBD   Aceto
   Ancazate BU   Anchor
   Butazate   Naugatuck SpA
   Butazate   Uniroyal
   Butazate 50D   Uniroyal
     (50 % slurry for use in latex)
   Butazin   Ticino
   Butyl Zimate   Vanderbilt
   Butyl Ziram   Pennwalt
   Butyl Ziram   Pennwalt
     (50 % aquous solution)
   Butyl Zirame   Prochim
   Eptac 4   Du Pont
   Eveite Butil Z   ACNA Montecatini
   JO 4013   Bozzetto
   Nocceler BZ   Ouchi Shinko
   Robac Z.B.U.D.   Robinson
   Sanceler BZ   Sanshin
   Soxinol BZ   Sumitomo
   Ultra Accelerator Di 13   Metallgesellschaft
   Vondac ZBUD   Vondelingenplaat
   Vulcafor ZNBC   ICI †
   ZDBC   Hasselt

4. Cream powder. M.W. 474.14; M.R. 95–108 °C.; S.G. 1.21.
   Non-staining ultra-acc. for NR, BR, SBR, NBR and their latices, acc. for EPDM; antidegr. for unvulcanized rubber and for non-staining grades of IIR; act. for thiazole- and other acid-type acc.
   △ Act. 28
   △ Antidegr. 103

66 1. Zinc dibutyldithiocarbamate - dibutylamine complex
    2. $[(C_4H_9)_2N.CS.S-]_2Zn . (C_4H_9)_2NH$
    3. Robac Z.B.U.D.X.   Robinson
    4. Golden brown liquid. M.W. 603.4.
    Non-staining acc. for solutions and for cements.

67 1. Zinc diethyldithiocarbamate
    2. $[(C_2H_5)_2N.CS.S-]_2Zn$

3. Accicure ZDC — Alkali
   Aceto ZDED — Aceto
   Ancazate ET — Anchor
   Etazin — Ticino
   Ethasan — Monsanto
   Ethazate — Naugatuck SpA
   Ethazate — Uniroyal
   Ethazate 50 D — Uniroyal
   (50 % slurry)
   Ethyl Zimate — Vanderbilt
   Ethyl Ziram — Pennwalt
   (oiled)
   Ethyl Ziram — Pennwalt
   (50 % aquous dispersion)
   Ethyl Zirame — Prochim
   Eveite Z — ACNA Montecatini
   Hermat ZDK — Dimitrova
   Nocceler EZ — Ouchi Shinko
   Robac ZDC — Robinson
   Sanceler EZ — Sanshin
   Soxinol EZ — Sumitomo
   Superaccelerator 1505 — Rhône Poulenc
   Superaccelerator 1505 N — Rhône Poulenc a
   Ultra Accelerator Di 7 — Metallgesellschaft
   Vondac ZDC — Vondelingenplaat
   Vulcafor ZDC — ICI
   Vulcafor ZDC — ICI (India)
   Vulkacit LDA — Bayer
   ZDEC — Hasselt
4. White powder. M.W. 361.91; M.P. 175 °C. min.; S.G. 1.49.
   Ultra-acc. for NR, SBR, NBR, IIR, and their latices; act. for
   thiazole-type acc. Non-staining.
   a. Special for use in latex.
   △ Act. 29

68 1. Zinc dimethyldithiocarbamate
2. $[(CH_3)_2N.CS.S-]_2Zn$

3. Aceto ZDMD — Aceto
   Ancazate ME — Anchor
   Cyazate M — Cyanamid
   Eptac 1 — Du Pont
   Eveite Metil Z — ACNA Montecatini
   Hermat ZDM — Dimitrova
   JO 4012 — Bozzetto
   JO 4012 Oleato — Bozzetto
   (83 % JO 4012,
   17 % paraffin oil)
   Metazin — Ticino
   Methasan — Monsanto
   Methazate — Naugatuck SpA
   Methazate — Uniroyal
   Methyl Zimate — Vanderbilt
   Methyl Ziram — Pennwalt
   (oiled and extruded)
   Methyl Ziram — Pennwalt
   (50 % dispersion)
   Methyl Zirame — Prochim
   Nocceler PZ — Ouchi Shinko
   Rhenocure ZMC — Rhein Chemie
   Robac ZMD — Robinson
   Sanceler PZ — Sanshin
   Soxinol PZ — Sumitomo
   Superaccelerator 1605 — Rhône Poulenc
   Ultra accelerator Di 4 — Metallgesellschaft
   Vondac ZMD — Vondelingenplaat
   Vulkacit L — Bayer
   ZDMC — Hasselt

4. White yellowish powder. M.W. 305.82; M.P. ca. 250 °C.; S.G. 1.66.

   Non-staining ultra-acc. for NR, SBR, IIR and their latices; act. for thiazole- and sulphenamide-type acc.

   △ Act. 30

69 1. Zinc dimethyl-pentamethylenedithiocarbamate

2. $[C_5H_8(CH_3)_2N.CS.S-]_2Zn$

3. Robac Z.L.   Robinson
4. Yellow-white powder. M.W. 442; M.P. 64 °C.
   Non-staining acc. for latex.

70 1. Zinc ethyl-phenyldithiocarbamate
       2. $[C_6H_5(C_2H_5)N.CS.S-]_2Zn$
       3. Ancazate EPH   Anchor
          Eveite P   ACNA Montecatini
          Hermat FEDK   Dimitrova
          Nocceler PX   Ouchi Shinko
          Sanceler PX   Sanshin
          Soxinol PX   Sumitomo
          Superaccelerator 1105   Rhône Poulenc
          Vondac ZEPD   Vondelingenplaat
          Vulcafor ZEP   ICI
          Vulkacit P extra N   Bayer
       4. White-yellow powder. M.W. 458.02; M.P. 205 °C.; S.G. 1.50.
          Ultra-acc. for NR, IIR, SBR and their latices.

71 1. Zinc N-pentamethylenedithiocarbamate
       2. $(C_5H_{10}N.CS.S-)_2Zn$
       3. Robac Z.P.D.   Robinson
          Vulkacit ZP   Bayer
          ZPMC   Hasselt
       4. Off-white powder. M.W. 385.9; M.P. 225 °C.; S.G. 1.60.
          Non-staining ultra-acc. for latex in combination with other zinc-
             dithiocarbamates; act. for thiazole- and sulpenamide-type acc.
          △ Act. 31

72 1. Zinc pentamethylenedithiocarbamate - piperidine complex
       2. $(C_5H_{10}N.CS.S-)_2Zn . C_5H_{10}NH$
       3. Robac Z.P.D.X.   Robinson
       4. White powder. M.W. 471.1; M.P. 140 °C.
          Also available as a 50 % paste.
          Acc. and act. for M.B.T.
          △ Act. 32

Thiuramsulphides

73 1. Cyclic thiuram-type products
2. —
3. Conac T   Du Pont a
     Sanceler CT   Sanshin b
4. a. Yellow powder. M.P. 96.1 °C.; S.G. 1.26.
       Acc. for SBR.
    b. Yellow powder. M.P. 115 °C. min.
       Del.act. acc. for SBR, NBR, IIR, IR, EPT.

74 1. Dimethyl-diphenyl-thiuram disulphide
2. $C_6H_5(CH_3)N.CS.S.S.CS.N(CH_3)C_6H_5$
3. DDTS   Bozzetto
     Vulkacit J   Bayer
4. White crystalline powder. M.W. 364; M.P. 175 °C. min.; S.G. 1.33.
   Non-staining del.act. acc. for NR, SBR, IR, BR, NBR. Sec. acc.
   for TMTS.

75 1. Dipentamethylene-thiuram disulphide
2. $(C_5H_{10}N.CS.S)_2$
3. Robac P.T.D.   Robinson a
     Robac P.T.D. 86   Robinson b
       (containing extra sulphur)
4. Cream coloured powder. M.W. 320.6;
    a. M.P. 120 °C.
    b. M.P. 110 °C.
   Non-staining acc. and vulc. for latex (gloves) and for IIR
    (pharmaceutical closures).
    △ Vulc. 32

76 1. Dipentamethylene-thiuram tetrasulphide
2. $(CH_2)_5N.CS.S_4.CS.N(CH_2)_5$
3. Accelerator 4P   Anchor
     DPTT   Hasselt
     Nocceler TRA   Ouchi Shinko
     Robac P.25   Robinson
     Soxinol TRA   Sumitomo

4. Light yellow powder or rods. M.W. 384.69; M.P. 115 °C.; S.G. 1.50.
   Ultra-acc. and vulc. for CSM, IIR, EPDM, NBR, SBR, IR, CR, NR; act. for thiazole- and sulphenamide-type acc.
   △ Act. 7
   △ Vulc. 34

77 1. Dipentamethylene-thiuram hexasulphide
   2.

$$H_2C \overset{\overset{\displaystyle CH_2-CH_2}{\diagup}}{\underset{\underset{\displaystyle CH_2-CH_2}{\diagdown}}{\hspace{2cm}}} N-\overset{\overset{\displaystyle S}{\|}}{C}-(S)_6-\overset{\overset{\displaystyle S}{\|}}{C}-N \overset{\overset{\displaystyle CH_2-CH_2}{\diagup}}{\underset{\underset{\displaystyle CH_2-CH_2}{\diagdown}}{\hspace{2cm}}} CH_2$$

   3. DPTT   Akron
      Sulfads   Vanderbilt
      Tetrone A   Du Pont
   4. Light gray powder. M.W. 448; M.P. 110 °C.; S.G. 1.53.
      Non-staining acc. for NR, SBR, NBR, CR, IR, IIR, EPDM, CSM. Also for latex. Vulc. for IR.
      △ Vulc. 33

78 1. Dipentamethylene-thiuram monosulphide
   2. $(C_5H_{10}N.CS)_2S$
   3. Robac P.T.M.   Robinson
   4. Yellow powder. M.W. 288.5; M.P. 110 °C.
      Del.act. acc. for chemically-blown cellular NR and SBR products; acc. for NBR in combination with MBTS.

79 1. Tetrabutyl-thiuram disulphide
   2. $[(C_4H_9)_2N.CS.S-]_2$
   3. Robac T.B.U.T.   Robinson
      Soxinol TBT   Sumitomo
      TBTS   Bozzetto
   4. Brown liquid. M.W. 408.7; solidifies at ca. 20 °C.; S.G. 1.1. Insoluble in water.
      Non-staining acc. in combination with P.T.D. or T.M.T. for NR, SBR, NBR in sulphurless compounds; ret. for CR; vulc. for NR, SBR, NBR.
      △ Ret. 10
      △ Vulc. 35

80 1. Tetrabutyl-thiuram monosulphide
  2. $(C_4H_9)_2N.CS.S.CS.N(C_4H_9)_2$
  3. Pentex   Uniroyal a
      Pentex Flour   Uniroyal b
        (Pentex 12.5 %, clay 87.5 %)
  4. a. Brown liquid. M.W. 376; S.G. 0.99.
          Del.action acc. for NR.
      b. Buff coloured powder. S.G. 2.16.
          Acc. for spongerubber.

81 1. Tetraethyl-thiuram disulphide
  2. $(C_2H_5)_2N.CS.S.S.CS.N(C_2H_5)_2$
  3. Accicure TET   Alkali
      Aceto TETD   Aceto
      Ancazide ET   Anchor
      Ethyl Thiram   Pennwalt
      Ethyl Thiurad   Monsanto
      Ethyl Tuads   Vanderbilt a
      Ethyl Tuex   Naugatuck SpA
      Ethyl Tuex   Uniroyal
      Etiurac   Ticino
      Eveite T   ACNA Montecatini
      Hermat TET   Dimitrova
      Nocceler TET   Ouchi Shinko
      Robac TET   Robinson
      Sanceler TET   Sanshin
      Soxinol TET   Sumitomo
      Super Accelerator 481   Rhône Poulenc
      TETD   Hasselt
      T.E.T.D.   Prochim
      TETS   Bozzetto
      Thiuram E   Du Pont
      Vondac TET   Vondelingenplaat
      Vulcafor TET   ICI
      Vulcafor TET   ICI (India)
  4. Grayish white powder and pellets. M.W. 296.54; M.P. 71–73 °C.;
        S.G. 1.26.
      a. Powder and rods.

Non-staining acc. for NR, SBR, NBR, BR, IIR, IR, EPDM; act. for
thiazole-, guanidine- and aldehyde-type acc.; vulc. for sulphurless compounds; stabilizer for Neoprene GN.

△ Act. 8
△ Antidegr. 96
△ Vulc. 36

82   1. Tetramethyl-thiuram monosulphide
        2. $(CH_3)_2N.CS.S.CS.N(CH_3)_2$
        3.

Aceto TMTM	Aceto
Ancazide IS	Anchor
Cyuram MS	Cyanamid
Eveite MST	ACNA Montecatini
Monex	Naugatuck SpA
Monex	Uniroyal
Mono Thiurad	Monsanto
Nocceler TS	Ouchi Shinko
Pennac MS	Pennwalt
Robac TMS	Robinson
Sanceler TS	Sanshin
Soxinol TS	Sumitomo
Superaccelerator 500	Rhône Poulenc
Thionex	Du Pont
TMTM	Hasselt
Unads	Vanderbilt
Vulcafor MS	ICI
Vulkacit Thiuram MS	Bayer
Vulkacit Thiuram MS/C (surface coated)	Bayer

        4. Yellow powder, pellets or rods. M.W. 208.37; M.R. 103–114 °C.;
            S.G. 1.37.
            Non-staining ultra-acc. for NR, NBR, IIR; act. for mercapto- and
            sulphenamide-type acc.
            △ Act. 10

83   1. Tetramethylthiuram disulphide
        2. $(CH_3)_2N.CS.S.S.CS.N(CH_3)_2$

Acc.

3. Accicure TMT	Alkali
Aceto TMTD	Aceto
Ancazide ME	Anchor
Cyuram DS	Cyanamid
Eveite 4MT	ACNA Montecatini
Hermat TMT	Dimitrova
Methyl Thiram (oiled and extruded)	Pennwalt
Methyl Tuads	Vanderbilt a
Metiurac	Ticino
Nocceler TT	Ouchi Shinko
RC Granulat TMTD (80 % TMTD, 20 % saturated hydrocarbons and dispersing agents)	Rhein Chemie
Robac TMT	Robinson
Sanceler TT	Sanshin
Soxinol TT	Sumitomo
Superaccelerator 501	Rhône Poulenc
Thiurad	Monsanto
Thiuram 16	Metallgesellschaft
Thiuram M	Du Pont
TMTD	Hasselt
TMTD	Akron
TMTD	Prochim
TMTS	Bozzetto
TMTS oleato (83 % TMTS, 17 % paraffin oil)	Bozzetto
Tuex	Naugatuck SpA
Tuex	Uniroyal
Vondac TMT	Vondelingenplaat
Vulcafor TMT	ICI
Vulcafor TMT	ICI (India)
Vulkacit Thiuram	Bayer
Vulkacit Thiuram C (surface coated)	Bayer
Vulkacit Thiuram GR (granules)	Bayer

4. White to yellow powder. M.W. 240.44; M.P. 146–148 °C.; S.G. 1.3–1.4.

 a. Powder and rods.

 Non-staining ultra-acc. and vulc.; act. for thiazole- and sulphen-amide-type acc.

 △ Act. 9

 △ Vulc. 37

84 1. Tetraethyl-thiuram disulphide / tetramethyl-thiuram disulphide blend

 2. —

 3. Methyl Ethyl Tuads   Vanderbilt a
    Pennac TM   Pennwalt b

 4. a. 50:50 mixture. White to cream rods. M.P. 62 °C. min.; S.G. 1.32.

    b. Light buff powder. M.W. 257.5; S.G. 1.24.

    Acc. and vulc. for sulphur-less or low-sulphur stocks of NR, SBR, EPDM.

    △ Vulc. 38

Heterocyclic compounds

85 1. 4,4'-Dithiodimorpholine

    syn. Dimorpholinyl disulphide

    Morpholine disulphide

 2.

 3. Deovulc M   Grandel
    Sulfasan R   Monsanto
    Vanax A   Vanderbilt
    Vondac DTDM   Vondelingenplaat
    Vulnoc R   Ouchi Shinko

 4. White to grey powder. M.W. 236.36; M.R. 115–127 °C.; S.G. 1.35. FDA appr.

    Vulc. and acc. for NR, SBR, NBR, IIR, EPDM.

    △ Vulc. 44

86 1. 2-Mercaptobenzimidazole

2.

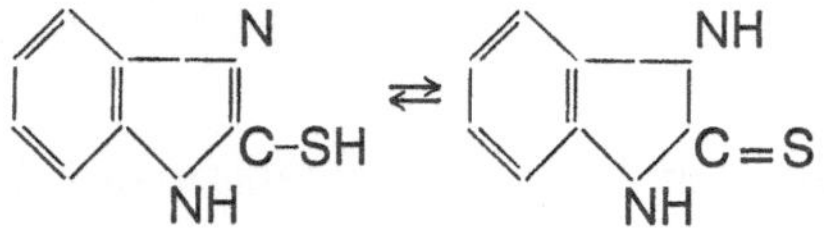

3.

Antigene MB	Sumitomo
Antioxidant MB	Dimitrova
Antioxidant MB	Bayer
Antivecchiante MB	ACNA Montecatini
MBI	Prochim
Nocrac MB	Ouchi Shinko
Permanax 21	Rhône Poulenc
Vondantox MBI	Vondelingenplaat

4. Yellow-white powder. M.W. 150; M.P. 290 °C. (with decomposition); S.G. 1.42.

Acc. for NR; non-staining oxi, heat, inh for NR, SBR, NBR; pept. for CR.

△ Antidegr. 94

△ Pept. 3

87 1. Carbon disulphide - 1,1'-methylene-dipiperidine reaction product

2. —

3. R.2 Crystals   Monsanto

4. Gray-white pellets. M.P. 55 °C. min.; S.G. 1.08–1.14.

Ultra-acc. for latex and fast-curing cements. FDA appr.

88 1. Thiohydropyrimidine

2. —

3. Thiate A   Vanderbilt

4. White crystalline powder. M.P. 250 °C. min.; S.G. 1.12.

Acc. for CR.

Thiazoles

89 1. Benzothiazyl-2-diethylsulphenamide

2.

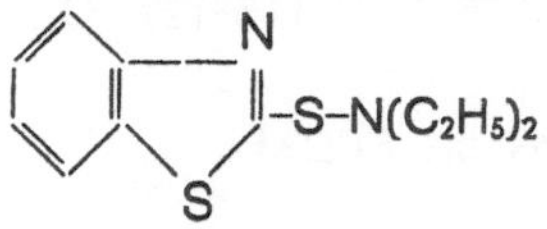

---

3. Vulkacit AZ   Bayer
4. Dark-brown oily liquid, hygroscopic. M.W. 238; S.G. 1.18.
   Del. action acc. for NR, IR, BR, SBR, NBR.

90 1. 1,3-Bis-(2-benzothiazyl mercaptomethyl) urea
        2.

3. El-Sixty   Monsanto
4. Buff powder. M.W. 418.63; M.P. 220 °C.; S.G. 1.35–1.41.
   Acc. for compounds containing large amounts of retarding fillers,
   requires ZnO and fatty acid; sec. acc. for ultra- and thiazole-
   type acc. in EPDM. Non-staining.

91 1. N-Cyclohexylamine salt of 2-mercaptobenzothiazole
        2.

3. Nocceler M-60   Ouchi Shinko
   Sanceler HM   Sanshin
4. Yellowish powder. M.W. 266.43; M.P. 150 °C. min.
   Ultra-acc. for NR, SBR, NBR. Also for latex.

92 1. 2,2'-Dibenzothiazyl disulphide
           syn. Di-(2-benzothiazyl) disulphide
        2.

3. Accicure MBTS   Alkali
   Altax   Vanderbilt
   Ancatax   Anchor
   Bowax AC/MBTS   Bozzetto
      (MBTS dispersed in Bowax C)
   Eveite DM   ACNA Montecatini
   MBTS   Akron

MBTS	Cyanamid
MBTS	Du Pont
MBTS	Naugatuck SpA
MBTS	Uniroyal
Mercasulf MBTS	Bozzetto
Nocceler DM	Ouchi Shinko
Pennac MBTS	Pennwalt
Pneumax DM	Dimitrova
Pneumax F	Dimitrova a
Rapid Accelerator 201	Rhône Poulenc
Sanceler DM	Sanshin
Soxinol DM	Sumitomo
Thiofide MBTS	Monsanto
Vulcafor MBTS	ICI
Vulcafor MBTS	ICI (India)
Vulkacit DM	Bayer

4. Yellowish powder. M.W. 332.50; M.P. 170–175 °C.; S.G. 1.50.

a. Mixture with basic acc.

Non-staining semi-ultra-acc. for NR, NBR, IIR, SBR. Also for latex. Ret. for CR.

△ Ret. 3

93 1. N,N-Dicyclohexyl-2-benzothiazyl sulphenamide

2.

3.

Nocceler DZ	Ouchi Shinko
Vulkacit DZ	Bayer

4. Coarse-grained brown powder. M.W. 346.58; M.P. 90 °C. min.; S.G. 1.20.

Del.action acc. for NR, IR, BR, SBR, NBR.

94 1. 2-(2,6-Dimethyl-4-morpholinyl mercapto) benzothiazole

2.

3. Santocure 26   Monsanto
4. Brown flakes. M.W. 280.42; M.P. 84 °C. min.; S.G. 1.23–1.29.
   Non-staining del.action acc. for natural and synthetic rubbers.

95 1. 2-(2,4-Dinitrophenyl) mercaptobenzothiazole
        syn. 2-(2,4-Dinitrophenyl) thiobenzothiazole
    2.

   3. Eveite 303   ACNA Montecatini
      Nocceler DBM   Ouchi Shinko
      Ureka Base   Monsanto
   4. Yellow powder. M.W. 333.35; M.P. 155 °C.; S.G. 1.61.
      Staining acc. for NR, IR, BR, SBR, NBR.

96 1. 2-Mercaptobenzothiazole
    2.

   3. Accicure MBT   Alkali
      Ancap   Anchor
      Captax   Vanderbilt
      Eveite M   ACNA Montecatini
      MBT   Akron
      MBT   Cyanamid
      MBT XXX   Cyanamid
        (very pure grade)
      MBT   Du Pont
      MBT   Naugatuck SpA
      MBT   Uniroyal
      Nocceler M   Ouchi Shinko

Pennac MBT	Pennwalt
Pneumax MBT	Dimitrova
Rapid Accelerator 200	Rhône Poulenc
Rotax	Vanderbilt
Sanceler M	Sanshin
Soxinol M	Sumitomo
Thiotax (MBT)	Monsanto
Vulcafor MBT	ICI
Vulcafor MBT	ICI (India)
Vulkacit Merkapto	Bayer

4. Light yellow powder. M.W. 167.25; M.R. 164–175 °C; S.G. 1.50. Non-staining semi-ultra acc. for NR, SBR, NBR, IIR. Also for latex. Ret. for CR; pept. for NR.

△ Pept. 4
△ Ret. 6

97 1. 2-Mercaptothiazoline

2.
$$H_2C-N$$
$$H_2C \quad C-SH$$
$$\diagdown S \diagup$$

3. Accelerator 2MT   Anchor
4. White powder. M.W. 119; M.P. 104–105 °C.

98 1. 4-Morpholinyl-2-benzothiazyl disulphide
syn. 2-(4-Morpholino-dithiobenzothiazole)

2.
$$CH_2-CH_2$$
$$\text{[benzothiazole]}-S-S-N \qquad O$$
$$CH_2-CH_2$$

3. Morfax   Vanderbilt
   Nocceler MDB   Ouchi Shinko
4. Cream powder. M.W. 284.35; M.R. 123–135 °C.; S.G. 1.51. Acc. for NR, SBR, BR, IIR, NBR, CR, EPDM.

99 1. Sodium 2-mercaptobenzothiazole

2.
$$\text{[benzothiazole]}-S-Na$$

3. Sanceler M-NA   Sanshin
4. Light-yellow crystalline powder. M.W. 189.24; M.P. 280 °C.
   Semi-ultra-acc. for NR latex.

100 1. Zinc-2-mercaptobenzothiazole
2.

3. Accicure ZMBT   Alkali
   Bantex   Monsanto
   Eveite MZ   ACNA Montecatini
   Hermat Zn MBT   Dimitrova
   Nocceler MZ   Ouchi Shinko
   OXAF   Naugatuck SpA
   OXAF   Uniroyal
   Pennac ZT   Pennwalt
   Pennac ZT-"W"   Pennwalt
     (with 10 % hydrocarbon)
   Rapid Accelerator 205   Rhône Poulenc
   Sanceler MZ   Sanshin
   Soxinol MZ   Sumitomo
   Vulcafor ZMBT   ICI
   Vulcafor ZMBT   ICI (India)
   Vulkacit ZM   Bayer
   Zenite (with hydrocarbon)   Du Pont
   Zenite Special   Du Pont
   Zetax   Vanderbilt
   Zinc Ancap   Anchor
   ZMBT   Cyanamid
   ZMBT waxed   Cyanamid
   ZMBT wettable   Cyanamid
4. Light cream powder. M.W. 397.9; M.P. 300 °C. (with decomposi-
     tion); S.G. 1.72.
   Acc. for NR-, SBR-, NBR-latices; non-staining oxi for latices.
   △ Antidegr. 99

Acc.

Sulphenamides

---

101 1. 2-(4-Morpholinyl-mercapto)-benzothiazole
   syn. N-Oxydiethylene-2-benzothiazole sulphenamide
      Benzothiazyl-2-sulphene morpholide

2.

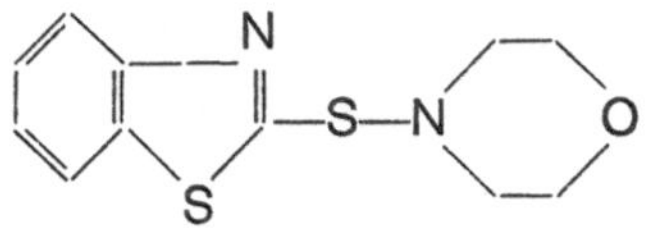

3.

Amax	Vanderbilt
Delac MOR	Uniroyal
NOBS Special	Cyanamid
Nocceler MSA	Ouchi Shinko
OBTS	Akron
Pennac MBS	Pennwalt
Sanceler NOB	Sanshin
Santocure MOR	Monsanto
Santocure MOR-90	Monsanto
(Santocure MOR/Thiofide 90 : 10)	
Soxinol NBS	Sumitomo
Vulcafor BSM	ICI
Vulcafor BSM	ICI (India)
Vulkacit MOZ	Bayer

4. Tan flakes. M.W. 252.3; M.R. 70–90 °C.; S.G. 1.37.
   Non-staining del.action acc. for NR, IR, BR, SBR, NBR. FDA
   appr.

102 1. N-tert. Butyl-benzothiazyl sulphenamide

2.

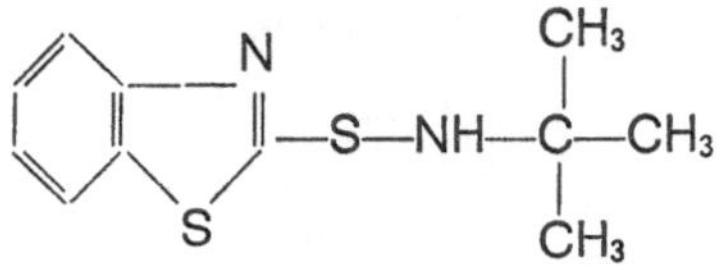

3.

BBTS	Akron
Conac NS	Du Pont
Delac NS	Uniroyal
Pennac TBBS	Pennwalt
Sanceler NS	Sanshin

Santocure NS	Monsanto
Santocure NS 50	Monsanto
(mixture with a pre-vulc. inhibitor)	
Soxinol NS	Sumitomo
Vulcafor BSB	ICI (India)
Vulkacit NZ	Bayer

4. Light-tan or buff powder or pellets. M.W. 238.37; M.P. 104 °C. min.; S.G. 1.26–1.32.
   Non-staining acc. for NR, SBR, IR, BR. FDA appr.

103   1. N-Cyclohexyl-2-benzothiazyl sulphenamide

2.

3.

Accicure HBS	Alkali
CBTS	Akron
Conac S	Du Pont
Cydac	Cyanamid
Delac S	Naugatuck SpA
Delac S	Uniroyal
Durax	Vanderbilt
Eveite MS	ACNA Montecatini
Nocceler CZ	Ouchi Shinko
Pennac CBS	Pennwalt
Rhodifax 16	Rhône Poulenc
Sanceler CM	Sanshin
Santocure	Monsanto
Soxinol CZ	Sumitomo
Sulfenax CB/30	Dimitrova
Vulcafor HBS	ICI
Vulcafor HBS	ICI (India)
Vulkacit CZ	Bayer

4. Greenish-tan flakes. M.W. 264.41; M.R. 90–108 °C.; S.G. 1.27–1.30.
   Del.action acc. for NR, SBR. Non-staining.

104 1. N,N-Diisopropyl-2-benzothiazyl sulphenamide

2.

3.

DIBS	Cyanamid
DIBS	Anchor
Dipac	Pennwalt
Nocceler PSA	Ouchi Shinko
Sanceler DIB	Sanshin

4. Light tan flakes. M.W. 266.42; M.R. 55–60 °C.; S.G. 1.21. Del.action acc. for NR, SBR, IR, BR.

Miscellaneous, mixtures and undisclosed compositions

105 1. Amine-salt of a dialkyl dithiophosphoric acid
2. —
3. Rhenocure AT   Rhein Chemie
4. White crystalline powder. M.R. 98–104 °C.; S.G. 1.04. Acc. for EPDM. Non-staining.

106 1. Bis-(2-ethylamino-4-diethylamino-triazine-6-yl) disulphide.

2.

3. Accelerator KA 9032   Bayer
4. Cream-coloured powder. M.W. 428; M.P. ca. 105 °C.; S.G. 1.126. Del.action acc. for NR, IR, SBR, NBR, BR.

107 1. N-Cyclohexyl-2-benzothiazyl sulphenamide / diphenylguanidine blend (70 : 30)
2. —
3. Pneumax CB/N   Dimitrova
4. Acc. for natural and synthetic rubbers.

108 1. 2,2'-Dibenzothiazyl disulphide / basic accelerators blend
2. —

3. Eveite F   ACNA Montecatini
   Vulkacit F   Bayer
   Vulkacit F/C   Bayer
     (surface coated)
4. Yellowish powder. M.P. 150 °C. min.; S.G. 1.31.
   Acc. for NR, IR, BR, SBR, NBR.

109 1. 2,2'-Dibenzothiazyl disulphide / diphenylguanidine blend
2. —
3. Accicure F   Alkali
   Vulcafor F   ICI
     (surface coated)
   Vulcafor F   ICI (India)
4. Cream-white powder. S.G. 1.46.

110 1. 2,2'-Dibenzothiazyl disulphide / hexamethylenetetramine blend
2. —
3. Nocceler Mix 3   Ouchi Shinko
   Soxinol G3   Sumitomo
4. Yellowish-white powder. M.P. 160 °C. min.
   Acc. for NR, SBR, NBR.

111 1. 2,2'-Dibenzothiazyl disulphide / 2-morpholinyl-thio-benzothiazyl
      blend (10 : 90)
2. —
3. Pennac MBS 90   Pennwalt
4. Tan chips. S.G. 1.36.

112 1. 2,2'-Dibenzothiazyl disulphide / N-oxydiethylene-2-benzothiazyl
      sulphenamide blend.
2. —
3. Amax No. 1   Vanderbilt
   NOBS No. 1   Cyanamid
     (90 : 10)
4. Tan flakes. M.R. 70–90 °C.; S.G. 1.40.
   Del.action acc. for use with high pH furnace blacks in NR and
   SBR.

113 1. 2,2'-Dibenzothiazyl disulphide / diphenylguanidine / hexamethy-lenetetramine blend.

2. —

3.
Accicure FN	Alkali
Nocceler F	Ouchi Shinko
Pneumax H/N	Dimitrova
Sanceler F	Sanshin
Soxinol F	Sumitomo
Vulcafor FN	ICI

4. Yellow powder. M.P. 145 °C.; S.G. 1.3–1.4.
Non-staining acc. for NR, SBR, IR, BR, NBR.

114 1. 2,2'-Dibenzothiazyl disulphide / thiuram disulphide blend.

2. —

3. Vulcafor DAU   ICI †

4. Cream powder. S.G. 1.47.

115 1. Dibutyl-xanthogen disulphide

2. $C_4H_9-O-\underset{\substack{\| \\ S}}{C}-S-S-\underset{\substack{\| \\ S}}{C}-O-C_4H_9$

3. C.P.B.   Uniroyal

4. Amber-coloured liquid. M.W. 298; S.G. 1.15.
Non-staining acc. for low-temperature vulcanization of NR, SBR, NBR, IIR, CR.

116 1. 2,4-Dinitrophenyl benzothiazyl sulphide / diphenylguanidine blend.

2. —

3. Pneumax U   Dimitrova

4. Yellow powder. M.P. 120 °C.
Acc. for natural and synthetic rubbers.

117 1. Dithiocarbamate / tetraethyl-thiuram disulphide blend (2 : 1)

2. —

3. Royalac 235   Uniroyal

4. Black powder. S.G. 1.46.
Acc. for EPDM.

118 1. Glycol-dimercaptoacetate
2. HSCH$_2$COO(CH$_2$)$_2$OCOCH$_2$SH
3. Robac G.D.M.A.   Robinson
4. Pale-yellow liquid. M.W. 210.3.
   Acc. for chlorobutylrubber. Non-staining.

119 1. Mercapto-derivatives / guanidines blends
2. —
3. Nocceler U   Ouchi Shinko
4. M.P. 125 °C.
   Acc. for NR, SBR, NBR.

120 1. 2,2'-Dibenzothiazyl disulphide / hexamethylenetetramine /
      2-mercaptobenzothiazole blend
2. —
3. Nocceler Mix 2   Ouchi Shinko
   Soxinol G2   Sumitomo
4. Light-yellow powder. M.P. 115 °C. min.
   Acc. for NR, SBR, NBR.

121 1. 2-Mercaptobenzothiazole / dithiocarbamate blend
2. —
3. Accicure DHC   Alkali
   Vulcafor DHC   ICI
   Vulcafor DHC   ICI (India)
4. Pale-yellow powder. S.G. 1.47.

122 1. 2-Mercaptobenzothiazole / ferric diethyldithiocarbamate blend
2. —
3. Nocceler EP-40   Ouchi Shinko
4. M.P. 140 °C.
   Acc. for EPDM.

123 1. 2-Mercaptobenzothiazole / ferric dimethyldithiocarbamate blend
2. —
3. Nocceler EP 10   Ouchi Shinko
4. M.P. 190 °C.
   Acc. for EPDM.

---

124 1. 2-Mercaptobenzothiazole / ferric dimethyldithiocarbamate / tetramethyl-thiuram disulphide blend

      2. —

      3. Nocceler EP 50   Ouchi Shinko

      4. M.P. 105 °C.
          Acc. for EPDM.

125 1. 2-Mercaptobenzothiazole / hexamethylenetetramine blend

      2. —

      3. Nocceler Mix 1   Ouchi Shinko
          Soxinol G 1   Sumitomo

      4. Light-yellow powder. M.P. 115 °C.
          Non-staining acc. for NR, SBR, NBR.

126 1. 2-Mercaptobenzothiazole / tetramethyl-thiuram disulphide blend

      2. —

      3. Captax Tuads blend (2 : 1)   Vanderbilt
          Herax N (2 : 1)   Dimitrova
          Nocceler 21   Ouchi Shinko
          Vulkacit MT/C (60 : 40)   Bayer
            (surface coated)

      4. Yellowish powder. M.R. 98–126 °C.; S.G. 1.42.
          Acc. for IIR.

127 1. 2-Mercaptobenzothiazole / zinc- N-diethyldithiocarbamate blend

      2. —

      3. Eveite 404   ACNA Montecatini
          Herax ZDKM/N   Dimitrova
          Nocceler EP-20   Ouchi Shinko
          Vulkacit MDA/C   Bayer
            (surface coated)

      4. Yellowish powder. M.P. 150–160 °C.; S.G. 1.44.
          Semi-ultra acc.

128 1. N-Cyclohexyl-2-benzothiazyl sulphenamide / tetramethyl-thiuram disulphide blend

      2. —

      3. Herax CB/X/N   Dimitrova

      4. Acc. for natural and synthetic rubbers.

129 1. 2,2'-Dibenzothiazyl disulphide / bis-(2-ethylamino-4-diethyl-
       amino-triazine-6-yl) sulphide / tetramethyl-thiuram disulphide
       blend (33 : 33 : 34)
    2. —
    3. Accelerator KA 9029   Bayer
    4. White-yellow powder. M.P. ca. 90 °C.; S.G. 1.4.
       Acc. for NR, IR, SBR, NBR, BR.

130 1. 2,2'-Dibenzothiazyl disulphide / bis-(2-ethylamino-4-diethyl-
       amino-triazine-6-yl) disulphide / tetramethyl-thiuram disulphide
       blend (46 : 46 : 8)
    2. —
    3. Accelerator KA 9030   Bayer
    4. Yellow-white powder. M.P. ca. 100 °C.; S.G. 1.11.
       Acc. for NR, IR, SBR, NBR, BR.

131 1. Tetramethyl-thiuram disulphide / zinc dimethyldithiocarbamate
       blend (50 : 50)
    2. —
    3. Eptac 2   Du Pont
       Nocceler EP-30   Ouchi Shinko
       Royalac 134   Uniroyal
    4. Off-white powder. S.G. 1.52.
       Non-staining acc. for EPDM, also for latex.

132 1. Bis-(2-ethylamino-4-diethylamino-triazine-6-yl) disulphide /
       tetramethyl-thiuram monosulphide blend (92 : 8)
    2. —
    3. Accelerator KA 9031   Bayer
    4. Yellow-white powder. M.P. ca. 100 °C.; S.G. 1.26.
       Acc. for NR, IR, SBR, NBR, BR.

133 1. Tetramethyl-thiuram monosulphide / zinc mercaptobenzothiazole
       blend (3 : 97)
    2. —
    3. Zenite A   Du Pont
       Zenite AM   Du Pont
         (micronized Zenite A)
    4. Cream-coloured powder. S.G. 1.53.
       Non-staining acc. for NR, SBR, NBR, IIR.

134 1. Thiazole / acid salt of diphenylguanidine blend
2. —
3. Vulcafor DAT   ICI a †
   Vulcafor DAW   ICI b †
4. a. Greyish powder. M.P. 127–130 °C.
   b. Yellow powder. S.G. 1.40.

135 1. Thiazole / guanidine blend
2. —
3. Blendac F   Anchor a
   Blendac FM   Anchor †
   Blendac V   Anchor b
   Ureka White   Monsanto c
4. Pale-cream powder. Non-staining.
   a. S.G. 1.43, acc. for NR, SBR.
   b. S.G. 1.46, acc. for NR, SBR, NBR.
   c. S.G. 1.39.

136 1. Thiazole / guanidine / hexamine blend
2. —
3. Blendac VN   Anchor
4. Cream powder. S.G. 1.47.
   Non-staining fast del.action acc.

137 1. Thiuram / thiazole blends
2. —
3. Accicure BT   Alkali a
   Blendac R   Anchor b
   Vulcafor BT   ICI (India) a
4. a. Cream-coloured powder. S.G. 1.44.
   b. Cream-coloured powder. S.G. 1.37.
   Acc. for NR, IIR.

138 1. Zinc o,o-di-n-butylphosphorodithioate
2. —
3. Vocol   Monsanto
   Vocol S   Monsanto
       (62 % active ingredient,
       38 % silicious carrier)

4. Clear yellow-green liquid. Flash Point 195 °C.; S.G. 1.25. Non-staining acc. for EPDM.

139 1. Zinc salt of dialkyldithiophosphoric acid.
2. —
3. Rhenocure TP   Rhein Chemie a
   Rhenocure TP/S   Rhein Chemie b
   (Mixture of Rhenocure TP
   and SiO_2, 67 : 33)
4. a. Greenish-yellow liquid. S.G. 1.25.
      Acc. for EPDM.
   b. White powder. S.G. 1.42.
      Acc. for EPDM.

140 1. Silicic acid / zinc salt of a thiophosphoricester blend
2. —
3. Deovulc S   Grandel
4. White powder. S.G. 1.4.
   Acc. for EPDM.

141 1. Composition undisclosed
2. —
3. Diak Super 6   Du Pont
   Eveite 202   ACNA Montecatini
   Harmon   Ticino
   Kenmix   Kenrich a
   Pennac NB liquid   Pennwalt
   Pennac NB powder   Pennwalt
      (74 % active ingredient on
      inert carrier)
   Pennac NB liquid   Vondelingenplaat
   Pennac powder   Vondelingenplaat
   Rapid Accelerator 465   Rhône Poulenc
   Rhodifax 7   Rhône Poulenc
   Rhodifax 10   Rhône Poulenc
   Rhodifax 14   Rhône Poulenc
   Robac 70   Robinson
   Robac Alpha   Robinson
   Robac C.S.   Robinson

Royalac 136		Uniroyal
Soxinol RL-13		Sumitomo
TA 11		Du Pont
Triox		Ticino
4. a.	△ Act. 44	
	△ Antidegr. 125	
	△ Vulc. 54	

Activators

1 1. Lead oxides
   2. —
   3. Mix Lpb 80   Bozzetto a
      Mix Pb 80   Bozzetto b
      Polyminium   Polychimie b
      Polytharge   Polychimie a
         grades A, B, D
      RC Granulat PbO   Rhein Chemie a
      RC Granulat Pb_3O_4 Rhein Chemie b
      —   Anchor c
   4. a. PbO dispersion. Acc. for CR; act. for NR, NBR, SBR, IIR;
            vulc. for CR, CSM.
      b. Pb_3O_4 dispersion. Acc. for CR; act. for IIR.
      c. PbO powder (litharge). Vulc. for CR.
      △ Acc. 1
      △ Vulc. 1

2 1. Magnesium oxide
   2. MgO
   3. Kenmag   Kenrich a
      Maglite   Merck b
         grades D, L, K, M, Y
      RC Granulat MgO   Rhein Chemie c
         (80 % MgO, 20 % saturated
         hydrocarbons and dispersion
         agents)
      Scorchguard   Anchor d
         grades C3, O, W
      Scorchguard O   Newalls e
      Struktol   Schill & Seilacher f
         grades WB 900, WB 902
         (coated)

4. M.W. 40.32.
   a. S.G. 2.02. Act. for CR.
   b. White powders. S.G. 3.3–3.5.
      Act. for SBR and fluoro-elastomers; vulc. for CR, CSM;
      antidegr. for CR, CSM, chlorobutyl, fluoro-elastomers,
      SBR.
   c. S.G. 2.06.
   d. Grade C3, powder, heavy calcined MgO.
      Grade O, putty, light calcined MgO.
      Grade W, powder, light calcined MgO.
   e. Dispersion. S.G. 2.08.
   f. Act. for CR and CSM mixtures.
   △ Antidegr. 1
   △ Vulc. 2

3 1. Magnesium oxide / zinc oxide blends
   2. —
   3. Struktol WB 890   Schill & Seilacher
   4. White powder, 45 % ZnO, 36 % MgO, 19 % dispersion agents.
      Act. for CR.

4 1. Siliciumdioxide / surface active agents blends
   2. —
   3. Aktivator 1987   Rhein Chemie a
      Aktivator 2009   Rhein Chemie a
      Aktivator 2642   Rhein Chemie a
      Aktivator R   Degussa b
   4. a. White powder. S.G. 1.5–1.75.
         Act. for vulcanizers and for blowing agents.
      b. Mixture of Ultrasil and hexanetriol.
         White powder.
      Acceleration-act. for SBR.

5 1. basic Zinccarbonate
   2. $ZnCO_3.2ZnO.3H_2O$
   3. Zinkoxid Transparent   Bayer
      —   Durham

4. White powder. S.G. 3.3–3.5.
   Act. for sulphur- and peroxide vulcanization of NR, IR, BR, SBR,
   NBR, IIR, CR, EPDM, CSM; vulc. for CR.
   △ Vulc. 7

6 1. Zinc oxide
   2. ZnO
   3.

Activox	Durham a
Decelox	Durham a
Durox 25	Durham a
Flocculent	Durham a
Loled	Durham a
Microx	Durham a
Mix Zn 60	Bozzetto b
Mix Zn 60 paste	Bozzetto c
Noled	Durham a
RC Granulat ZnO	Rhein Chemie d
(80 % ZnO, 20 % saturated hydrocarbons and dispersion agents)	
RC Zinkoxid 64	Rhein Chemie d
(90 % ZnO, 10 % dispersion agents)	
S.R.Q.	Durham a
Struktol Neopast	Schill & Seilacher e
Struktol WB 700	Schill & Seilacher f
Zinkoxid Aktiv	Bayer a
—	ACNA Montecatini a
—	Anchor g

   4. a. White powder. M.W. 81. S.G. 5.4.
      b. 60 % micronized ZnO in microcrystalline waxes and hydro-
         carbon resins. White flakes. M.P. 50 °C., S.G. 1.7–1.8.
      c. White paste. M.P. 64–68 °C., S.G. 1.85–1.9.
      d. Yellow-white powder. S.G. 4.0.
      e. Grey paste. Act. for CR- and CSM mixtures. S.G. 2.1.
      f. White powder. Act. for CR- and CSM mixtures.
      g. Act. for CR.

Act.

ORGANIC

Thiuramsulphides

7 1. Dipentamethylene-thiuram tetrasulphide
   2. $(CH_2)_5N.CS.S_4.CS.N(CH_2)_5$
   3. Accelerator 4P   Anchor
      DPTT   Hasselt
      Nocceler TRA   Ouchi Shinko
      Robac P.25   Robinson
      Soxinol TRA   Sumitomo
   4. Light yellow powder or rods. M.W. 384.69; M.P. 115 °C.; S.G. 1.50.
      Act. for thiazole- and sulphenamide-type acc.; ultra-acc. and vulc. for CSM, IIR, EPDM, NBR, SBR, IR, CR, NR.
      △ Acc. 76
      △ Vulc. 34

8 1. Tetraethyl-thiuram disulphide
   2. $(C_2H_5)_2.N.CS.S.S.CS.N(C_2H_5)_2$
   3. Accicure TET   Alkali
      Aceto TETD   Aceto
      Ancazide ET   Anchor
      Ethyl Thiram, extruded   Pennwalt
      Ethyl Thiurad   Monsanto
      Ethyl Tuads   Vanderbilt a
      Ethyl Tuex   Naugatuck SpA
      Ethyl Tuex   Uniroyal
      Etiurac   Ticino
      Eveite T   ACNA Montecatini
      Hermat TET   Dimitrova
      Nocceler TET   Ouchi Shinko
      Robac TET   Robinson
      Sanceler TET   Sanshin
      Soxinol TET   Sumitomo
      Superaccelerator 481   Rhône Poulenc
      TETD   Prochim
      TETD   Hasselt
      TETS   Bozzetto

Thiuram E	Du Pont
Vondac TET	Vondelingenplaat
Vulcafor TET	ICI
Vulcafor TET	ICI (India)

4. Grayish white powders and pellets. M.W. 296.54; M.P. 71–73 °C.; S.G. 1.26.

   a. Powder and rods.

   Act. for thiazole-, guanidine- and aldehyde-type acc.; acc. for NR, SBR, NBR, BR, IIR, IR, EPDM; vulc. for sulphurless compounds; stabilizer for Neoprene GN. Non-staining.

   △ Acc. 81

   △ Antidegr. 96

   △ Vulc. 36

9 1. Tetramethyl-thiuram disulphide

    2. $(CH_3)_2N.CS.S.S.CS.N(CH_3)_2$

    3.

Accicure TMT	Alkali
Aceto TMTD	Aceto
Ancazide ME	Anchor
Cyuram DS	Cyanamid
Eveite 4MT	ACNA Montecatini
Hermat TMT	Dimitrova
Methyl Thiram oiled	Pennwalt
Methyl Thiram extruded	Pennwalt
Methyl Tuads	Vanderbilt a
Metiurac	Ticino
Nocceler TT	Ouchi Shinko
RC Granulat TMTD (80 % TMTD, 20 % saturated hydrocarbons and dispersion agents)	Rhein Chemie
Robac TMT	Robinson
Sanceler TT	Sanshin
Soxinol TT	Sumitomo
Superaccelerator 501	Rhône Poulenc
Thiurad	Monsanto
Thiuram 16	Metallgesellschaft
Thiuram M	Du Pont

TMTD	Akron
TMTD	Hasselt
TMTD	Prochim
TMTS	Bozzetto
TMTS oleato	Bozzetto
(83 % TMTS, 17 % paraffin oil)	
Tuex	Naugatuck SpA
Tuex	Uniroyal
Vondac TMT	Vondelingenplaat
Vulcafor TMT	ICI
Vulcafor TMT	ICI (India)
Vulkacit Thiuram	Bayer
Vulkacit Thiuram C	Bayer
(surface coated)	
Vulkacit Thiuram GR	Bayer
(granules)	

4. White to yellow powder. M.W. 240.44; M.P. 146–148 °C.; S.G. 1.3–1.4.

a. Powder and rods.

Act. for thiazole- and sulphenamide-type acc.; non-staining ultra-acc. and vulc.

△ Acc. 83

△ Vulc. 37

10 1. Tetramethyl-thiuram monosulphide

2. $(CH_3)_2N.CS.S.CS.N.(CH_3)_2$

3.

Aceto TMTM	Aceto
Ancazide IS	Anchor
Cyuram MS	Cyanamid
Eveite MST	ACNA Montecatini
Monex	Naugatuck SpA
Monex	Uniroyal
Mono Thiurad	Monsanto
Nocceler TS	Ouchi Shinko
Pennac MS	Pennwalt
Robac TMS	Robinson
Sanceler TS	Sanshin
Soxinol TS	Sumitomo

Superaccelerator 500	Rhône Poulenc
Thionex	Du Pont
TMTM	Hasselt
Unads	Vanderbilt
Vulcafor MS	ICI
Vulkacit Thiuram MS	Bayer
Vulkacit Thiuram MS/C (surface coated)	Bayer

4. Yellow powder, pellets or rods. M.W. 208.37; M.R. 103–114 °C.; S.G. 1.37.

   Act. for mercapto- and sulphenamide-type acc.; non-staining ultra-acc. for NR, NBR, IIR.

   △ Acc. 82

Amines

11 1. Butyraldehyde - aniline condensation product

   2. —

   3.

Accelerator 21	Anchor
Antox Special	Du Pont
Beutene	Uniroyal
Butanyl 1	Ticino
Nocceler 8	Ouchi Shinko
Rapid Accelerator 300A	Rhône Poulenc
Vulcafor BA	ICI

   4. Orange-red oily liquid. S.G. 0.94–1.02.

   Act. for thiazoles, thiurams and guanidines; semi-ultra acc. for Neoprene W and for hydrocarbon rubber; antidegr. for CR.

   △ Acc. 14

   △ Antidegr. 41

12 1. Cyclohexyl ethyl amine

   2. $(CH_2)_5CH.NH.C_2H_5$

   3. Vulkacit HX   Bayer

   4. Pale yellow liquid. M.W. 127; B.P. 165 °C.; S.G. 0.85.

   Sec. acc. and act

   △ Acc. 7

13 1. Dibenzylamine / monobenzylamine blend
   2. —
   3. D.B.A.   Uniroyal
   4. Yellow liquid. S.G. 1.03.
      Act. for dibutyl-xanthogen disulphide acc.

14 1. Heptaldehyde - aniline condensation product
   2. —
   3. Heptene Base   Uniroyal
   4. Dark brown liquid. S.G. 0.92.
      Act. for thiazole- and thiuram-type acc.; acc. for NR.
      △ Acc. 17

15 1. Hexamethylenetetramine (with additives)
   2. $(CH_2)_6N_4$
   3. Aceto HMT   Aceto
      Eveite UR   ACNA Montecatini
      Herax UTS   Dimitrova
      Hexa K   Degussa
      Hexalit   Chemko
        grades S, 01, 02
      Nocceler H   Ouchi Shinko
      RC Granulat Hexa   Rhein Chemie
        (80 % hexa and 20 %
        saturated hydrocarbons
      and dispersion agents)
      Rhenocure Hexa   Rhein Chemie †
      Sanceler H   Sanshin
      Soxinol H   Sumitomo
      Vulkacit H 30   Bayer
   4. Hygroscopic white powder. M.W. 140.19; sublimates at 263 °C.;
      S.G. 1.3.
      Act. for mercapto-, sulphenamide-, thiuram- and zinc dithio-
      carbamate-type acc.; acc. for NR, SBR, NBR.
      △ Acc. 2

16 1. Triethanolamine
   2. $N(CH_2CH_2OH)_3$

3. Ankoltet   Anchor
   TEA   Shell
       grades TEA 85 %,
       TEA Commercial, TEA 98 %
4. Water-white liquid. M.W. 149.19; B.P. 286 °C.; S.G. 1.12.
   Act. for accelerators.

Dithiocarbamates

17   1. 2-Benzothiazyl-N,N-diethyldithiocarbamate
       syn. 2-Benzothiazyl-N,N-diethylthiocarbamyl sulphide
      2.

      3. Ethylac   Pennwalt
        Ethylac   Vondelingenplaat
        Nocceler 64   Ouchi Shinko
      4. Yellow powder. M.W. 282.45; M.P. ca. 70 °C.; S.G. 1.27.
        Non-staining acc. and act. for NR, BR, SBR, NBR, IR.
        △ Acc. 45

18   1. Bismuth dimethyldithiocarbamate
      2. $[(CH_3)_2N.CS.S-]_3Bi$
      3. BDMC   Hasselt
        Bismate   Vanderbilt
        JO 6000   Bozzetto
        Robac Bi.D.D.   Robinson
      4. Yellow powder and pellets. M.W. 569.66; M.P. above 227 °C.
        (with decomposition); S.G. 2.02.
        Act. for thiazole- and sulphenamide-type acc.; acc. for NR, SBR,
        IIR.
        △ Acc. 41

19   1. Copper dimethyldithiocarbamate
      2. $[(CH_3)_2N.CS.S-]_2Cu$
      3. CDMA   Hasselt
        Cumate   Vanderbilt
        Hermat Cu   Dimitrova

JO 4015	Bozzetto
Nocceler TTCU	Ouchi Shinko
Robac Cu.D.D.	Robinson
Sanceler TTCU	Sanshin
Soxinol MK	Sumitomo

4. Dark brown powder. M.W. 303.98; M.P. above 325 °C. (with decomposition); S.G. 1.75.
Act. for thiazole- and sulphenamide-type acc.; acc. for SBR, IIR.
△ Acc. 44

20
1. 2,4-Dinitrophenyl-dimethyldithiocarbamate
2. $(CH_3)_2N.CS.S.C_6H_3(NO_2)_2$–2,4
3. Safex   Uniroyal
4. Yellow crystalline powder. M.W. 287; M.R. 140–145 °C.; S.G. 1.57.
Act. for thiazole-type acc.; del. action acc.
△ Acc. 50

21
1. Lead dimethyldithiocarbamate
2. $[(CH_3)_2N.CS.S–]_2Pb$
3.

JO 4014	Bozzetto
LDMC	Hasselt
Ledate	Vanderbilt
Robac L.M.D.	Robinson

4. White powder. M.W. 447.65; M.P. above 310 °C.; S.G. 2.43.
Act. for thiazole- and sulphenamide-type acc.; acc. for NR, BR, IR, SBR, IIR.
△ Acc. 52

22
1. Piperidinium-pentamethylenedithiocarbamate
syn. N-Pentamethyleneammonium-N-pentamethylenedithio-carbamate
2. $(CH_2)_5N.CS.S.H_2N(CH_2)_5$
3.

Accelerator 552	Du Pont
Accelerator 2P	Anchor
Nocceler PPD	Ouchi Shinko
Pentalidine	Prochim
Robac P.P.D.	Robinson
Vulkacit P	Bayer

4. Cream powder. M.W. 246; M.P. 175 °C.; S.G. 1.19.
   Act. for thiuram- and thiazole-type acc.; acc. for NR, NBR, SBR;
   pept. for Neoprene G- and KNR types.
   △ Acc. 54
   △ Pept. 6

23 1. Selenium diethyldithiocarbamate
   2. [(C₂H₅)₂N.CS.S–]₄Se

3. Ethyl Selenac	Vanderbilt
Ethyl Seleram SA-66-1 (oiled)	Pennwalt
Seleniame	Prochim
Soxinol SE	Sumitomo

   4. Yellow-orange powder. M.W. 672; M.R. 59–85 °C.; S.G. 1.32.
      Act. for thiazole-type acc.; acc. for IIR, vulc. for NR, NBR, SBR.
      △ Acc. 57
      △ Vulc. 40

24 1. Sodium dibutyldithiocarbamate
   2. (C₄H₉)₂N.CS.S.Na

3. Ancazate WSB (48 % aquous solution)	Anchor
Butyl Soderame (47 % aquous solution)	Prochim
Nocceler TP	Ouchi Shinko
Robac SBUD (45 % aquous solution)	Robinson
SBTC (40% aquous solution)	Bozzetto
Soxinol TP (40% aquous solution)	Sumitomo
Tepidone (47 % aquous solution)	Du Pont

   4. Clear brown liquid. M.W. 227; S.G. 1.075–1.09.
      Act. for thiazole-type acc.; ultra-acc. for latices and for reclaim-
      ed mixes.
      △ Acc. 59

25 1. Sodium diethyldithiocarbamate
   2. $(C_2H_5)_2N.CS.S.Na$
   3.

Ethyl Soderame	Prochim a
Eveite L	ACNA Montecatini b
Nocceler SDC	Ouchi Shinko a
Pennac SDED	Pennwalt
(25 % aquous solution)	
Robac S.E.D.	Robinson
(23 % aquous solution)	
Sanceler ES	Sanshin
(20–22 % aquous solution)	
Soxinol ESL	Sumitomo
(18–22 % aquous solution)	
Super Accelerator 1500	Rhône Poulenc c
Vondac SDED	Vondelingenplaat
(25 % aquous solution)	
Vulcafor SDC	ICI †

   4. a. White crystalline powder. M.W. 171.26; S.G. 1.3.
      b. White crystalline powder. M.W. 207.29; M.P. 90–92 °C.; S.G.
         1.30.
      c. Hydrate. M.W. 225; S.G. 1.30.
      Act. for guanidine-type acc.; ultra-acc. for NR- and SBR latices.
      Non-staining.
      △ Acc. 60

26 1. Tellurium diethyldithiocarbamate
   2. $[(C_2H_5)_2N.CS.S-]_4Te$
   3.

Soxinol TE	Sumitomo
TDEC	Hasselt
Tellurac	Vanderbilt
Tellurame	Prochim

   4. Orange-yellow powder and rods. M.W. 720.69; M.R. 108–118 °C.;
      S.G. 1.44.
      Act. for thiazole- and thiuram-type acc.; acc. for NR, SBR, NBR,
      IIR, EPDM.
      △ Acc. 62

27 1. Zinc dibenzyldithiocarbamate
   2. $[(C_6H_5CH_2)_2N.CS.S-]_2Zn$

3. Arazate   Uniroyal a
    Robac Z.B.E.D.   Robinson b
    ZBEC   Hasselt b
4. Creamy-white powder. M.W. 610.2; S.G. 1.41.
    a. M.R. 160–175 °C.
    b. M.P. 182 °C. min.
    Act. for thiazole- and sulphenamide-type acc.; ultra-acc. for IIR,
       SBR, NR, also for latex.
    △ Acc. 64

28   1. Zinc dibutyldithiocarbamate
      2. [(C$_4$H$_9$)$_2$N.CS.S–]$_2$Zn
      3.

Aceto ZDBD	Aceto
Ancazate BU	Anchor
Butazate	Naugatuck SpA
Butazate	Uniroyal
Butazate 50D	Uniroyal
(50 % slurry for use in latex)	
Butazin	Ticino
Butyl Zimate	Vanderbilt
Butyl Ziram	Pennwalt
Butyl Ziram	Pennwalt
(50 % aquous solution)	
Butyl Zirame	Prochim
Eptac 4	Du Pont
Eveite Butil Z	ACNA Montecatini
JO 4013	Bozzetto
Nocceler BZ	Ouchi Shinko
Robac Z.B.U.D.	Robinson
Sanceler BZ	Sanshin
Soxinol BZ	Sumitomo
Ultra Accelerator Di 13	Metallgesellschaft
Vondac ZBUD	Vondelingenplaat
Vulcafor ZNBC	ICI †
ZDBC	Hasselt

      4. Cream powder. M.W. 474.14; M.R. 95–108 °C.; S.G. 1.21.
         Act. for thiazole- and other acid-type acc.; non-staining ultra-
           acc. for NR, BR, SBR, NBR and their latices, acc. for EPDM;

antidegr. for unvulcanized rubber, and for non-staining grades
of IIR.

△ Acc. 65

△ Antidegr. 103

29 1. Zinc diethyldithiocarbamate
   2. $[(C_2H_5)_2N.CS.S-]_2Zn$
   3.

Accicure ZDC	Alkali
Aceto ZDED	Aceto
Ancazate ET	Anchor
Etazin	Ticino
Ethasan	Monsanto
Ethazate	Naugatuck SpA
Ethazate	Uniroyal
Ethazate 50D	Uniroyal
(50 % dispersion)	
Ethyl Zimate	Vanderbilt
Ethyl Ziram (oiled)	Pennwalt
Ethyl Ziram	Pennwalt
(50 % dispersion)	
Ethyl Zirame	Prochim
Eveite Z	ACNA Montecatini
Hermat ZDK	Dimitrova
Nocceler EZ	Ouchi Shinko
Robac ZDC	Robinson
Sanceler EZ	Sanshin
Soxinol EZ	Sumitomo
Superaccelerator 1505	Rhône Poulenc
Superaccelerator 1505(N)	Rhône Poulenc a
Ultra Accelerator Di 7	Metallgesellschaft
Vondac ZDC	Vondelingenplaat
Vulcafor ZDC	ICI
Vulcafor ZDC	ICI (India)
Vulkacit LDA	Bayer
ZDEC	Hasselt

   4. White powder. M.W. 361.91; M.P. 175 °C. min.; S.G. 1.49.
      Act. for thiazole-type acc.; non-staining ultra-acc. for NR, SBR,
      NBR, IIR and their latices (a. special for use in latex).
      △ Acc. 67

30 Zinc dimethyldithiocarbamate
2. [(CH$_3$)$_2$N.CS.S–]$_2$Zn

3.

Aceto ZDMD	Aceto
Ancazate ME	Anchor
Cyzate M	Cyanamid
Eptac 1	Du Pont
Eveite Metil Z	ACNA Montecatini
Hermat ZDM	Dimitrova
JO 4012	Bozzetto
JO 4012 oleato	Bozzetto
(83 % JO 4012, 17 % paraffin oil)	
Metazin	Ticino
Methasan	Monsanto
Methazate	Naugatuck SpA
Methazate	Uniroyal
Methyl Zimate	Vanderbilt
Methyl Ziram	Pennwalt
grades extruded, oiled, 50 % dispersion	
Methyl Zirame	Prochim
Nocceler PZ	Ouchi Shinko
Rhenocure ZMC	Rhein Chemie
Robac ZMD	Robinson
Sanceler PZ	Sanshin
Soxinol PZ	Sumitomo
Superaccelerator 1605	Rhône Poulenc
Ultra accelerator Di 4	Metallgesellschaft
Vondac ZMD	Vondelingenplaat
Vulkacit L	Bayer
ZDMC	Hasselt

4. White-yellowisch powder. M.W. 305.82; M.P. ca. 250 °C.; S.G. 1.66.

Act. for thiazole- and sulphenamide-type acc.; non-staining ultra-acc. for NR, SBR, IIR and their latices.

△ Acc. 68

31 1. Zinc N-pentamethylenedithiocarbamate
2. (C$_5$H$_{10}$N.CS.S–)$_2$Zn

3. Robac Z.P.D.   Robinson
   Vulkacit ZP   Bayer
   ZPMC   Hasselt
4. Off-white powder. M.W. 385.9; M.P. 225 °C.; S.G. 1.60.
   Act. for thiazole- and sulphenamide-type acc.; non-staining ultra-acc. for latex in combination with other zinc-dithiocarbamates.
   △ Acc. 71

32
1. Zinc pentamethylenedithiocarbamate - piperidine complex
2. $(C_5H_{10}N.CS.S-)_2Zn$. $C_5H_{10}NH$
3. Robac Z.P.D.X.   Robinson
4. White powder. M.W. 471.1; M.P. 140 °C. Also available as a 50 % paste.
   Acc. and act. for M.B.T.
   △ Acc. 72

Miscellaneous, mixtures and undisclosed compositions

33
1. Diarylguanidine blend
2. $R_1.NH.C(:NH).NH.R_2$
3. Accelerator 49   Cyanamid
4. White to pinkish-white powder. M.P. ca. 134 °C.; S.G. 1.20.
   Act. for MBT or MBTS in SBR; acc. for NR.
   △ Acc. 30

34
1. Dibutylammoniumoleate
2.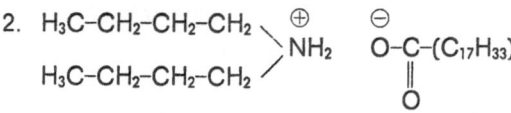
3. Activator 1102   Anchor
   Barak   Du Pont
   DOB   Organo Synthèse
4. Dark-amber liquid. M.W. 409; Flash Point 102 °C.; S.G. 0.88.
   Act. for thiazole-, thiuram- and sulphenamide-type acc.; act.-ret. for thiuram-type acc.
   △ Ret. 4

35 1. N,N'-Dibutylthiourea
2. $(C_4H_9NH)_2CS$
3.

Accelerator DBT	BASF
DBTU	Prochim
—	Degussa
Pennzone B	Pennwalt
Pennzone B	Vondelingenplaat
Robac DBTU	Robinson

4. Off-white powder. M.W. 188.3; M.P. 65 °C.
Act. for EPDM and NR; acc. for mercaptan-modified CR; anti-degr. for NR-latex and for thermoplastic SBR.
△ Acc. 22
△ Antidegr. 109

36 1. N,N'-Diphenylguanidine
2. $HN = C(NH.C_6H_5)_2$
3.

Accicure DPG	Alkali
Denax	Dimitrova
Denax DPG	Vychodočeské
DPG	Cyanamid
DPG	Anchor
DPG	Monsanto
DPG	Rhône Poulenc
Eveite D	ACNA Montecatini
Nocceler D	Ouchi Shinko
Pennac DPG	Pennwalt
Sanceler D	Sanshin
Soxinol D	Sumitomo
Vulcafor DPG (surface coated)	ICI
Vulcafor DPG	ICI (India)
Vulkacit D	Bayer

4. White crystalline non-hygroscopic powder or paste. M.W. 211.26; M.R. 144–146 °C.; S.G. 1.19.
Medium acc. for use with thiazoles and sulphenamides.
△ Acc. 31

37 1. Sym. Diphenyl-thiourea
syn. Thiocarbanilide

2. $C_6H_5.NH.CS.NH.C_6H_5$

3.

A-L Thiocarbanilide	Monsanto
DPTU	Prochim
Eveite TC	ACNA Montecatini
Nocceler C	Ouchi Shinko
Soxinol C	Sumitomo
Stabilisator C	Bayer
(formerly Vulkacit CA)	
Vulcafor TC	ICI
—	Degussa

4. Cream-white powder. M.W. 228; M.P. 149 °C. min.; S.G. 1.31.
   Non-staining sec. acc. for CR, EPDM.
   △ Acc. 25

38 1. N,N'-Di-ortho-tolylguanidine

2. $2–CH_3.C_6H_4.NH.C(:NH)NH.C_6H_4.CH_3–2$

3.

DOTG	Anchor
DOTG	Cyanamid
DOTG	Du Pont
DOTG	Rhône Poulenc
Eveite DOTG	ACNA Montecatini
Nocceler DT	Ouchi Shinko
Soxinol DT	Sumitomo
Vulcafor DOTG	ICI
Vulcafor DOTG	ICI
(surface coated)	
Vulkacit DOTG	Bayer
Vulkacit DOTG/C	Bayer
(surface coated)	

4. White powder. M.W. 239; M.R. 167–173 °C.; S.G. 1.19.
   Act. for acidic and neutral acc.; slow-curing acc. for NR, SBR,
   NBR.
   △ Acc. 32

39 1. Polyoxyethyleneglycol
   syn. Polyethyleneglycol

2. $H–(O–CH_2–CH_2)_n–OH$

3. Glicogum 4000   Bozzetto a
   PEG   Shell b
      grades 200, 300, 400, 555M,
      600, 800, 1000, 1500, 4000,
      4000F, 4000P, 6000
4. a. White flakes. M.R. 54–57 °C.; S.G. 1.12.
   b. The numerical suffix is an indication of the average molecular weight.
      PEG 200, 300, 400 colourless liquids. PEG 555M, 600 soft white materials. PEG 800, 1000, 1500, 4000, 6000 white waxy solids.
   Act. for NR, SBR.

40 1. ortho-Tolylbiguanide
    2. $H_2N–C(:NH)–NH–C(:NH)–C_6H_4.CH_3$
    3. Accelerator 80   Du Pont
       Eveite 1000   ACNA Montecatini
       Nocceler BG   Ouchi Shinko
       Vulkacit 1000   Bayer
       Vulkacit 1000/C   Bayer
         (surface coated)
    4. White powder. M.W. 191.24; M.P. 140 °C.; S.G. 1.2.
       Act. for zinc dithiocarbamates, thiuram-, mercapto- and sulphenamide-type acc.; acc. for NR, IR, BR, SBR, NBR.
       △ Acc. 34

41 1. Triallylcyanurate
    2. $(CH_2:CHCH_2–OC:N–)_3$

    3. Aktivator OC   Bayer
       Aktivator OC   Degussa
    4. Viscous, pasty or crystalline, depending on the temperature. M.W. 249; M.P. 29 °C.; B.P. 149–150 °C. (at 2 mm Hg); S.G. 1.14.
       Act. for peroxide-vulcanization.

42 1. Modified urea
    2. —
    3. Activator 1203   Anchor a
       Activator DN   Du Pont b

Attivante 4030	Bozzetto a
Attivante 4030 pasta	Bozzetto a
BIK	Naugatuck SpA b
BIK	Uniroyal b
BK	Ouchi Shinko b
RIA NC	Nat. Polychemicals a

4. White powders, S.G. ca. 1.3.
   a. act. for thiazole-, thiuram- and sulphenamide type acc.
   b. act. for nitrogeneous blowing agents.

43 1. Zinc salts of a mixture of fatty acids in which lauric acid predominates.

    2.

$$CH_3-(CH_2)_{10}-\overset{\overset{\displaystyle O}{\|}}{C}-O-Zn-O-\overset{\overset{\displaystyle O}{\|}}{C}-(CH_2)_{10}-CH_3$$

    3.

Laurex	Naugatuck SpA
Laurex	Uniroyal

    4. Yellowish waxy powder. M.P. 95–105 °C.; S.G. 1.15.
       Non-staining act. for NR, SBR, NBR, CR.

44 1. Composition undisclosed
    2. —
    3.

Additiv-Paste 1100	Rhein Chemie
Additiv-Paste 1600	Rhein Chemie
Aktiplast	Rhein Chemie
Aktiplast T	Rhein Chemie
Aktivator 3555	Rhein Chemie
Aktivator B	Rhein Chemie
Kenmix	Kenrich a
Ridacto	Kenrich
Ridacto 75	Kenrich
(75 % dispersion)	
Ritardante 01	Bozzetto
Silacto	Kenrich
Silacto 75	Kenrich
(75 % dispersion)	

    4. a. △ Acc. 141
          △ Antidegr. 125
          △ Vulc. 54

Antidegradants

Antioxidant	oxi
Antiozonant	ozo
Antiflexcracking-agent	flex
Heat-stabilizer	heat
Metalpoison-inhibitor	inh

INORGANIC

1   1. Magnesium oxide
    2. MgO
    3. Kenmag   Kenrich a
       Maglite   Merck b
         grades D, L, K, M, Y
       RC Granulat MgO   Rhein Chemie c
         (80 % MgO, 20 % saturated
         hydrocarbons and dispersion
         agents)
       Scorchguard   Anchor d
         grades C3, O, W
       Scorchguard O   Newalls e
       Struktol   Schill & Seilacher f
         grades WB 900, WB 902
         (coated)
    4. M.W. 40.32.
      a. S.G. 2.02. Act. for CR.
      b. White powders. S.G. 3.3–3.5.
         Antidegr. for CR, CSM, chlorobutyl, SBR, fluoro-elastomers;
           act. for SBR, and fluoro-elastomers; vulc. for CR, CSM.
      c. S.G. 2.06.
      d. Grade C3, powder, heavy calcined MgO. Grade O, putty,
         light calcined MgO. Grade W, powder, light calcined MgO.

e. Dispersion. S.G. 2.08.

f. Act. for CR and CSM mixtures.

△ Act. 2

△ Vulc. 2

ORGANIC

Hydrocarbons and waxes

2	1. Waxes and paraffinic products	
	2. —	
	3. AC Polyethylene	Allied
	grades 6, 6A, 7, 8, 8A, 615, 617, 617A, x1702, G 201.	
	Antilux	Rhein Chemie
	grades 540, 550, 600, 654, AO, AOL, L.	
	Antisun	Ross
	Aristowax 165	Ross
	Controzon	Grandel
	grades ASM, Plus, T, W.	
	Heliozone	Du Pont
	Heliozone Special	Du Pont
	Lichtschutzwachs	Schlickum
	grades 85R, 102, 116, 38, 335M.	
	Ozonschutzwachs 110	Bayer
	Ozonschutzwachs 111	Bayer
	Ozonschutzwachs 110	Rhein Chemie
	Ozonschutzwachs 111	Rhein Chemie
	Ozonschutzwachs DOG	Grandel
	RC Lichtschutzmittel 520	Rhein Chemie
	Shellwax	Shell
	grades 100, 200, 400.	
	Sunnoc	Ouchi Shinko
	Sunnoc N	Ouchi Shinko
	Sunproof	Naugatuck SpA
	grades 100, Regular, Improved, Super.	

Sunproof   Uniroyal
    grades Extra, Super, Improved,
    Regular, Junior, 713.

Sunproofing Wax   Ross
    grades 1343, 3920.

4. —

Phenols

3   1. Alkylated p-cresol
      2. —
      3. Naugawhite 434   Uniroyal
      4. Viscous yellow liquid. S.G. 1.08.
         Oxi for dry NR and for NR- and SBR latex.

4   1. Alkylated bisphenol
      2. —
      3. Antigene NW   Sumitomo a
         Cyanox 53   Cyanamid b
           (on inert carrier)
         Naugawhite   Naugatuck SpA a
         Naugawhite powder   Naugatuck SpA b
           (on inert carrier)
         Naugawhite   Rubber Regenerating a
         Naugawhite   Uniroyal a
         Naugawhite powder   Uniroyal b
           (on inert carrier)
      4. a. Slightly viscous liquid. S.G. 0.96.
         b. White powder. S.G. 1.19.
         Non-staining oxi for NR, SBR, CR, IR, NBR, BR.

5   1. Alkylated phenols blends
      2. —
      3. Antioxidant KSM   Bayer
         Antioxidant KSM-EM-33   Bayer
         Arrconox AHT   Rubber Regenerating

Arrconox AHT powder	Rubber Regenerating
(on inert carrier)	
Cyanox LF	Cyanamid
Cyanox LF	Anchor
Oxystop	Arwal
grades 320, 330.	
Wingstay T	Goodyear

4. Non-staining powders and liquids.
Oxi for natural and synthetic rubber.

6
1. Alkylated styrenated phenol
2. —
3. Wingstay V   Goodyear
4. Amber liquid. S.G. 1.00. FDA appr.
Non-staining oxi for natural and synthetic rubber. Also for latex.

7
1. Alkylated thio-bisphenol
2. —
3. Antioxidant 423   Uniroyal
(form. trade-name VBUI)
4. White crystalline solid. M.P. 115–117 °C.; S.G. 1.07.
Non-staining oxi for rubber.

8
1. Alkyl- and aryl-substituted phenol blend
2. —
3. Antioxidant DS   Bayer a
(form. Antioxidant KA 9013)
Antioxidant DS/F   Bayer b
Antioxidant TSP   Bayer c
4. a. Yellow-reddish viscous liquid. B.P. 150 °C. min. (10 Torr); S.G. 0.915.
Non-staining oxi, heat, flex for NR, SBR, IR, BR, NBR, CR.
b. White powder with inorganic filler, 50 % active ingredient.
Non-staining oxi, heat, flex for mixtures of NR with SBR, IR, BR, NBR, CR.
c. Viscous orange-red liquid. B.P. 170–180 °C. (10 Torr).

9
1. Alkylphenol sulphide
2. —

3. Rhenadox APS   Rhein Chemie
4. White crystalline powder. M.P. 72 °C. min.; S.G. 1.06.
   Non-staining oxi, heat.

10 1. 1,1-Bis-(4-hydroxy phenyl) cyclohexane
     2.

$$CH_2-CH_2$$
H_2C C —OH
$$CH_2-CH_2$$ —OH

3. Antigene W   Sumitomo
4. White powder. M.W. 268; M.P. 175 °C. min.
   Non-staining oxi, heat for NR, SBR.

11 1. 6-tert Butyl, 2,4-dimethyl phenol
     2.

OH
$(CH_3)_3C$—   —CH_3

CH_3

3. Antioxidant 624   Raschig
4. Yellow liquid. M.W. 178.26; B.P. 250 °C.; S.G. 0.95–0.96.
   Oxi.

12 1. 6-tert Butyl, 3-methyl phenol derivatives
     2. —
     3. Antigene WL–L   Sumitomo
        Antigene WL–O   Sumitomo
     4. Reddish-brown tacky liquid.
        Non-staining oxi for NR and for synthetic rubber.

13 1. 2,2'-Butylidene-bis-(6-tert butyl p-cresol)
     2.

CH_3 CH_3
$(CH_3)_3C$—   —C_4H_8—   —$C(CH_3)_3$

OH   OH

3. Anullex PBH   Pearson

4. Off-white crystalline powder. M.W. 382.24; M.P. 127 °C.; S.G. 0.56.

Non-staining oxi for dry rubber and latex.

14 1. 4,4'-Butylidene-bis-(2-tert butyl 5-methylphenol)
syn. 4,4'-Butylidene-bis-(6-tert butyl m-cresol)

2.

$(CH_3)_3C$— HO— —C_4H_8— —$C(CH_3)_3$ OH
CH_3 H_3C

3.

Antigene BBM	Sumitomo
Anullex PBA 15	Pearson
Santowhite Powder	Monsanto
Sumilizer BBM	Sumitomo

4. White crystalline powder. M.W. 382.56; M.P. 209 °C.; S.G. 1.03–1.09.

Non-staining oxi for natural and synthetic rubber, also for latex.
FDA appr.

15 1. 2,6-Di-tert butyl-p-cresol
syn. 2,6-Di-tert butyl-4-methyl phenol
2,6-Di-tert butyl hydroxy-toluene

2. $C_6H_2(OH)[(CH_3)_3C-]_2(CH_3)-1-2,6-4$

3.

Antigene BHT	Sumitomo
Antioxidant 4	Dimitrova
Antioxidant DBPC	Organo Synthèse
grades AT 1, AP 3, AP 4, AB 2.	
Antioxidant KB	Bayer
Anullex BHT	Pearson
Anullex BHT techn. grade 1	Pearson
B.H.T.	Bennett
B.H.T.	Raschig
Bisoxol 220	CdF
Catalin Antioxidant	Ashland
grades CAO-1, CAO-3.	
DTBP	Raschig
Imbutol	Metallgesellschaft
Imbutol E	Metallgesellschaft
(50 % aquous solution)	

Ionol CP	Shell
Narox	Naugatuck SpA
Naugard BHT	Uniroyal
Nocrac 200	Ouchi Shinko
Nonox TBC	ICI
(only available in Great Britain)	
Rhenadox DB	Rhein Chemie
Tenamene 3	Eastman

4. White to yellow flakes. M.W. 220; M.P. 68.9 °C.; S.G. 1.03.
Non-staining oxi, heat, flex for NR, IR. Also for latex. In combination with waxes ozo.

16 1. 4,4'-Thio-bis-(6-tert butyl-m-cresol)
syn. 4,4'-Thio-bis-(6-tert butyl-3 methyl phenol)
4,4'-Thio-bis-(2-tert butyl-5 methyl phenol)
Di-(3-tert butyl-4-hydroxy-6-methyl phenyl) sulphide

2.

3.

Antigene WX	Sumitomo
Antigene WX–R	Sumitomo
(pure grade)	
Antioxidant TBM-6	Organo Synthèse
T grade	
Anullex PSA 10	Pearson
Nocrac 300	Ouchi Shinko
Santowhite Crystals	Monsanto
Sumilizer WX	Sumitomo
Sumilizer WX–R	Sumitomo
(pure grade)	

4. White to greyish-white powder. M.W. 358.55; M.P. 150 °C. min.;
S.G. 1.06–1.12.
Non-staining oxi and heat for natural and synthetic rubber,
especially for CR, also for latex. FDA appr.

17 1. 3,5-Di-tert butyl-4-hydroxy-toluene

2.

$(CH_3)_2C-$⟨ring, OH top, CH_3 bottom⟩$-C(CH_3)_2$

3. Antigene BHT   Sumitomo
4. White powder or flakes. M.W. 220; M.P. 69 °C. min.
   Non-staining oxi for NR, NBR, IR, PE, also for latex; inh for NR.

18 1. 4,4'-Dihydroxydiphenyl
    2. $HOC_6H_4C_6H_4OH$
    3. Antioxidant DOD   Bayer
    4. Gray powder. M.W. 186; M.P. 260 °C.; S.G. 1.37.
       Non-staining oxi, heat for NR dry rubber, and in combination
       with di-beta-naphthyl-p-phenylenediamine for NR latex.

19 1. N,N'-Disalicylidene-1,2-propanediamine
    2.

3. Copper Inhibitor 50   Du Pont
4. Reddish-brown liquid. M.W. 270; Flash-Point 36 °C.; S.G. 0.99.
   Inh for NR, SBR, CR. Also for latex.

20 1. 1,6-Hexanediol bis-[3-(3',5'-di-tert butyl-4'-hydroxyphenyl)
       propionate]
    2.

3. Irganox 259   Ciba Geigy
4. White crystalline powder. M.W. 638.9; M.P. 104–109 °C.
   Oxi.

21 1. hindered Bisphenol
    2. —

3. Antioxidant NKF   Bayer
4. White crystalline powder. M.P. 155 °C. min.; S.G. 1.12.
   Oxi, heat, flex. Inh in combination with Antioxidant MB.
   (2-Mercapto-benzimidazole).

22 1. hindered Phenols
    2. —
    3. Agerite Geltrol   Vanderbilt a
       Antioxidant NV 1   Bayer a
       Antioxidant 431   Uniroyal a
       Antioxidant 555   Pitt Consol a
       Endox 11T   Akron b
       Nevastain Powder 2170   Neville b
       Santowhite 54   Monsanto a
       Santowhite 54–S   Monsanto b
          (on inert carrier)
       Wingstay L   Goodyear b
       Zalba Special   Du Pont b
    4. a. liquid
       b. powder.
       Oxi, flex.

23 1. 2,2'-Methylene-bis-(4-ethyl-6-tert butylphenol)
    2. $CH_2[-(2)-C_6H_2(OH)C_2H_5\{C(CH_3)_3\}-1,4,6]_2$
    3. Antioxidant 425   Cyanamid
       Antioxidant 425   Anchor
       Antioxidant TBE 9   Organo Synthèse
       Endox 22   Akron
       Naruxol 25   Naugatuck SpA
    4. Cream to white powder. M.W. 368.54; M.R. 119–125 °C.; S.G.
       1.10.
       Oxi, heat, flex for NR, SBR, NBR, CR. Also for latex.

24 1. 2,2'-Methylene-bis-(4-methyl-6-tert butylphenol)
    2.

3. Antigene MDP   Sumitomo
   Antioxidant 2246   Cyanamid
   Antioxidant 2246   Anchor
   Antioxidant BKF   Bayer
   Bisoxol O   CdF
   Catalin Antioxidant   Ashland
   grades CAO-5, CAO-14.

   Endox 21T   Akron
   Naruxol 15   Naugatuck SpA
   Nocrac NS-6   Ouchi Shinko
   Sumilizer MDP   Sumitomo
   Synox 5LT   Neville Synthèse
   Synox 5P   Neville Synthèse
   (pure grade)

4. Colourless to light-cream crystalline powder. M.W. 340.51; M.P. 120 °C, min.; S.G. 1.08.
   Non-staining oxi, heat, flex, inh for NR, BR, SBR, NBR. Also for latex.

25 1. 2,2'-Methylene-bis-(dimethyl-4,6-phenol)
   2.

   3. Permanax 28 "H.V."   Rhône Poulenc
   4. Oiled powder, cream coloured. M.W. 256; M.P. 120 °C.; S.G. 1.1.
      Oxi, heat, flex for NR, SBR. Also for latex.

26 1. 2,2-Methylene-bis-[6-(alpha methyl cyclohexyl)-p-cresol]
   2.

   3. Nonox WSP   ICI
      Nonox WSP   ICI America

---

4. Off-white crystalline powder. M.W. 420; M.P. 130 °C.; S.G. 1.17.
   Non-staining oxi for synthetic rubber and for polyolefins.

27 1. 2,2'-Methylene-bis-(4-methyl-6-cyclohexylphenol)
2. $CH_2-[-(2)-C_6H_2.(OH).CH_3.C_6H_{13}-1,4,6]_2$
3. Antioxidant ZKF   Bayer
4. White crystalline powder. M.W. 380; M.P. 118 °C.; S.G. 1.08.
   Oxi, heat, flex, inh for NR, IR, BR, SBR, NBR.

28 1. 2,2'-Methylene-bis-(tert nonyl-cresol)
2. $CH_2-[-(2)-C_6H_2.(OH).CH_3.C_9H_{19}]_2$
3. Anox G 1   Bozzetto a
   Anox G 1 polvere   Bozzetto b
      (absorbed on molecular sieve)
4. a. Viscous amber-coloured liquid. S.G. 0.96.
   b. Light brown powder. S.G. 1.01.
   Non-staining oxi, heat for NR, SBR, CR.

29 1. Aryl-alkyl phenol blends
2. —
3. Permanax 24   Rhône Poulenc
4. Yellow liquid. S.G. 1.079 and cream powder (60 % Permanax 24
   and 40 % mineral powder).
   Non-staining oxi, heat for NR, SBR, NBR, IR, CR, BR.

30 1. Phenol-derivatives blends
2. —
3. Antioxidant MP   Anchor a
   Nonox HO   ICI b
4. a. Amber liquid. B.P. 250 °C.; S.G. 0.95.
   b. Amber liquid. S.G. 0.94.
   Oxi, flex.

31 1. Polybutylated bisphenol-A blend
2. —
3. Agerite Superlite   Vanderbilt a
   Agerite Superlite solid   Vanderbilt b
      (70 % active ingredient)

4. a. Amber liquid. S.G. 0.965.
   b. Gray-tan solid. S.G. 1.26.
   Oxi.

32 1. Nickel salt of 3,5-di-tert butyl-4-hydroxybenzyl-phosphonic acid-monoethylester

2.

$$\left[\; HO{-}\!\!\underset{C(CH_3)_3}{\overset{C(CH_3)_3}{\bigcirc}}\!\!{-}CH_2{-}\overset{\overset{O}{\uparrow}}{\underset{OC_2H_5}{P}}{-}O^{\ominus} \; \right]_2 \; Ni \quad 2\oplus$$

3. Irgastab 2002   Ciba Geigy
4. Pale-yellow powder. M.W. 713.47.
   Oxi, light stabilizer.

33 1. Octadecyl-3-(3',5'-di-tert butyl-4'-hydroxyphenyl) propionate

2.

$$HO{-}\!\!\underset{C(CH_3)_3}{\overset{C(CH_3)_3}{\bigcirc}}\!\!{-}CH_2CH_2{-}CO{-}O{-}C_{18}H_{37}$$

3. Irganox 1076   Ciba Geigy
4. White crystalline powder. M.W. 530.9; M.P. 49–54 °C.
   Oxi.

34 1. Pentaerythritol-tetrakis-[3-(3',5'-di-tert butyl-4'-hydroxyphenyl) propionate]

2.

$$\left[\; HO{-}\!\!\underset{C(CH_3)_3}{\overset{C(CH_3)_3}{\bigcirc}}\!\!{-}CH_2CH_2{-}CO{-}O{-}CH_2{-} \; \right]_4 \; C$$

3. Irganox 1010   Ciba Geigy
4. White crystalline powder. M.W. 1177.7; M.P. 110–115 °C.
   Oxi.

35 1. 1,1-Bis-(4-hydroxy phenyl) cyclohexane - organic amine reaction product

2. —
3. Antigene WA   Sumitomo
4. Light brown powder. M.P. 80 °C. min.
   Non-staining oxi, ozo, heat for NR, SBR, NBR, CR.

36 1. 6-tert Butyl-m-cresol - sulphur dichloride reaction product
2. —
3. Santowhite MK   Monsanto
4. Dark-brown viscous liquid. M.P. 20–35 °C.; S.G. 1.03–1.09.
   Non-staining oxi for dry rubber and latex. FDA appr.

37 1. Styrenated phenols
2.

3. Accinox SP   Alkali a
   Agerite SPAR   Vanderbilt
   Anox G2   Bozzetto
   Anox G2 Polvere   Bozzetto
      (absorbed on molecular sieves)
   Antigene S   Sumitomo
   Antioxidant SP   Anchor
   Arrconox SP   Rubber Regenerating
   Arrconox SP Powder   Rubber Regenerating
      (dispersed in an inert powder
      base)
   Montaclere   Monsanto
   Montaclere SE   Monsanto
      (self-emulsifying grade)
   Nocrac SP   Ouchi Shinko
   Nonox SP   ICI a
   Nonox SP   ICI (India) a
   Stabilite SP   Reichhold
   Wingstay S   Goodyear
4. Light-yellowish transparent gluey liquid.
   Non-staining oxi, ozo, heat for NR, SBR, NBR, CR. FDA appr.
   a. Mixture of styrenated phenols.

38 1. 4,4'-Thio-bis-(di-sec amyl phenol)
2. $C_{32}H_{50}O_2S$
3. Santowhite L   Monsanto
4. Dark viscous liquid. M.W. 508.8; S.G. 0.94–1.04.
Non-staining oxi for latex. FDA appr.

39 1. 1,3,5-Trimethyl-2,4,6-tris-(3,5-di tert butyl-4-hydroxybenzyl)
benzene
2.

3. Ionox 330   Shell
4. Near-white crystalline solid. M.W. 768; M.P. 244 °C.
Oxi.

Derivatives of aniline and of substituted aniline

40 1. Acetaldehyde-aniline condensation product
2. —
3. Crylene   Uniroyal a
   (mixture with stearic acid
   67 : 33)
Nocceler K   Ouchi Shinko b
VGB   Uniroyal c
Vulcafor RN   ICI †
4. a. Thick brown paste. S.G. 1.014.
   b. Reddish-brown powder. M.P. 55 °C.

c. Brown resinous powder. M.R. 60–80 °C,; S.G. 1.15.
Staining oxi, heat for NR; acc. for NR.
△ Acc. 11

41 1. Butyraldehyde - aniline condensation product
2. —
3. Accelerator 21   Anchor
Antox Special   Du Pont
Beutene   Uniroyal
Butanyl-1   Ticino
Nocceler 8   Ouchi Shinko
Rapid Accelerator 300A   Rhône Poulenc
Vulcafor BA   ICI
4. Orange-red oily liquid. S.G. 0.94–1.02.
Antidegr. for CR; semi-ultra acc. for Neoprene W and hydro-
carbon rubber; act. for thiazoles, thiurams and guanidines.
△ Acc. 14
△ Act. 11

42 1. 4,4'-Diamino diphenyl methane
2. $CH_2(C_6H_4.NH_2)_2$
3. Robac 4.4   Robinson a
Tonox   Uniroyal b
4. a. Light brown powder. M.W. 198.3; M.R. 75–85 °C.
b. Brown waxy lump. S.G. 1.18.
Antifrosting agent for NR; acc. for CR; ret. for IIR.
△ Acc. 15
△ Ret. 2

43 1. N,N-Di-(N'-ethylidene-anilino) aminobenzene
2.
$$C_6H_5-NH-HC-\underset{\underset{CH_3}{|}}{\overset{\overset{C_6H_5}{|}}{N}}-CH-NH-C_6H_5$$
$$\underset{CH_3}{|}\underset{CH_3}{|}$$
3. Eveite A   ACNA Montecatini
4. Brown oily liquid. M.W. 331.45; S.G. 1.06–1.07.
Antidegr. and acc.
△ Acc. 16

44 1. Arylamines blends
2. —
3.

Accinox HFN	Alkali
Anox Beta	Bozzetto
Anox JO	Bozzetto
Nonox HFN	ICI
Nonox HFN	ICI (India)
Nonox HFN	ICIANZ

4. Gray flakes, granules, pellets. S.G. 1.18–1.22.
Staining oxi, heat, flex, for NR, SBR, NBR, CR and for isobuty-lene-isoprene copolymers.

45 1. N-Methyl, N,4-dinitroso-aniline / inert clay blend
2. —
3. Heat Pro   Conestoga
4. Antidegr.

Derivatives of phenylene-diamine

46 1. Alkyl-aryl p-phenylenediamines
2. —
3.

Aceto Ozone	Aceto a
Flexzone 5L	Uniroyal b
Santoflex 134	Monsanto c
Santoflex 134 SE	Monsanto c
(self-emulsifying grade)	

4. a. Blend of alkyl-aryl p-phenylenediamine with waxes. Tan granulated beads. M.P. 107.2 °C.; S.G. 0.98.
Oxi, ozo, flex, inh, heat for NR, SBR.
   b. Dark-brown crystalline solid.
Staining oxi, ozo for SBR, IR, BR. (Formerly Antiozonant 437)
   c. Dark oil. Crystallization Point <18.3 °C.; S.G. 0.993.
Oxi for SBR.

47 1. N,N'-Bis-(1-ethyl, 3-methyl pentyl)-p-phenylenediamine
syn. N,N'-Di-(3,5-methyl heptyl)-p-phenylenediamine
2. $(C_2H_5.CHCH_3.CH_2.CHC_2H_5.NH)_2C_6H_4$

3. Antozite 2 — Vanderbilt
   Flexzone 8L — Uniroyal
   Santoflex 17 — Monsanto
   UOP 88 — UOP
4. Dark red-brown liquid. M.W. 332.56; S.G. 0.87–0.93.
   Staining ozo, flex, for natural and synthetic rubber.

48
1. N,N'-Bis-(1-methyl heptyl)-p-phenylenediamine
   syn. N,N'-Di-(2-octyl)-p-phenylenediamine
2.

$$C_6H_{13}-\underset{\underset{CH_3}{|}}{\overset{\overset{H}{|}}{C}}-\overset{\overset{H}{|}}{N}-\langle=\rangle-\overset{\overset{H}{|}}{N}-\underset{\underset{CH_3}{|}}{\overset{\overset{H}{|}}{C}}-C_6H_{13}$$

3. Antozite 1 — Vanderbilt
   ANTO$_3$"B" — Pennwalt
   ANTO$_3$"D" — Pennwalt
      (isomeric)
   Santoflex 217 — Monsanto
   UOP 288 — UOP
4. Dark red-brown liquid. M.W. 332.56; S.G. 0.87–0.93.
   Staining oxi for NR.

49
1. N,N'-Bis-(1,4-dimethyl pentyl)-p-phenylenediamine
2.

$$CH_3-\underset{\underset{H}{|}}{\overset{\overset{CH_3}{|}}{CH}}-CH_2-CH_2-\overset{\overset{CH_3}{|}}{CH}-\overset{\overset{}{|}}{N}-\langle=\rangle-\overset{\overset{}{|}}{N}-\overset{\overset{CH_3}{|}}{CH}-CH_2-CH_2-\overset{\overset{CH_3}{|}}{CH}-CH_3$$

3. Antioxidant 4030 — Bayer
   Antozite MPD — Vanderbilt
   Cyzone DH — Cyanamid
   Eastozone 33 — Eastman
   Flexzone 4L — Uniroyal
   Santoflex 77 — Monsanto
   UOP 788 — UOP
4. Reddish-brown liquid. M.W. 304.51; S.G. 0.894–0.906.
   Staining general purpose ozo.

50
1. Diaryl-p-phenylenediamine (mixed)
2. —

3. Akroflex AZ   Du Pont a
   Antigene DTP   Sumitomo b
   Wingstay 100   Goodyear c
   Wingstay 200   Goodyear d
4. a. Brown powder. M.P. 103 °C.; S.G. 1.27.
      Staining ozo for SBR, CR.
   b. Black powder.
      Oxi, ozo, heat, flex for NR, SBR, NBR.
   c. Blue-brown solid. M.P. 90–105 °C.; S.G. 1.20.
      Oxi, ozo, flex.
   d. Brown-black semi-solid. M.P. 50–60 °C.; S.G. 1.20.
      Oxi, ozo, flex.

51 1. N,N'-Di-cyclohexyl-p-phenylenediamine
2. $C_6H_{11}.NH.C_6H_4.NH.C_6H_{11}$
3. UOP 26   UOP
4. Brown flakes. M.W. 272.4; M.P. 102–108 °C.; S.G. 0.59.
   Staining ozo, flex for NR, NBR, CR, IR, BR.

52 1. N,N'-Di-heptyl-p-phenylenediamine
2. $C_7H_{15}.NH.C_6H_4.NH.C_7H_{15}$
3. ANTO$_3$"G"   Pennwalt
4. Reddish-brown liquid. M.W. 304; S.G. 0.898.
   Staining ozo for NR, SBR.

53 1. N-(1,3-Dimethyl butyl)-N'-phenyl-p-phenylenediamine
2. $CH_3.CHCH_3.CH_2.CHCH_3.NH.C_6H_4.NH.C_6H_5$
3. Antioxidant 4020   Bayer
   Antozite 67   Vanderbilt
   Antozite 67 S   Vanderbilt
     (50 % active ingredient)
   Flexzone 7L   Uniroyal
   Nonox ZC   ICI
   Santoflex 13   Monsanto
   Santoflex 13 S   Monsanto
     (Santoflex / Carbon black 1 : 1)
   UOP 588   UOP

4. Dark coloured crystals or flakes. M.W. 268.39; M.R. 40–44 °C.;
   S.G. 0.9–1.1.
   Staining general purpose oxi, ozo, flex, inh.

54 1. N,N'-Di-beta-naphthyl-p-phenylenediamine
  2.

  3. Aceto DIPP   Aceto
     Agerite White   Vanderbilt
     Agerite White   Anchor
       (Antioxidant 123 outside U.K.)
     Antigene F   Sumitomo
     Antioxidant DNP   Bayer
     Antivecchiante DNP   ACNA Montecatini
     Nocrac White   Ouchi Shinko
     Nonox Cl   ICI
     Nonox Cl   ICIANZ
     Santowhite Cl   Monsanto
  4. Grey powder. M.W. 360; M.P. 230 °C.; S.G. 1.28.
     Staining oxi, heat, inh for NR, SBR, NBR. Also for latex.

55 1. N,N'-Diphenyl-p-phenylenediamine / phenyl-alpha-naphthylamine
         blend
  2. —
  3. Akroflex C   Du Pont a
       (65 : 35)
     Nocrac 500   Ouchi Shinko
  4. a. Dark-gray pellets. M.P. 75 °C.; S.G. 1.23.
     Staining oxi, ozo, flex, heat for NR, SBR, CR, IIR.

56 1. N,N'-Diphenyl-p-phenylenediamine
  2. $C_6H_5.NH.C_6H_4.NH.C_6H_5$
  3. Agerite DPPD   Vanderbilt
     Antigene P   Sumitomo
     Antioxidant DPPD   Anchor
     DPPD   Monsanto
     Inibitore OB   ACNA Montecatini a
     J–Z–F   Uniroyal

Nocrac DP   Ouchi Shinko
Nonox DPPD   ICI
Permanax 18   Rhône Poulenc

4. Light gray powder. M.W. 260; M.P. 140 °C.; S.G. 1.22.
   a. M.P. 125–130 °C.
   Staining oxi, ozo, flex, heat for NR, SBR, NBR, IR, BR, CR.

57
1. N-Isopropyl-N'-phenyl-p-phenylenediamine
   syn. 4-Isopropylamino-diphenylamine
2. $C_3H_7.NH.C_6H_4.NH.C_6H_5$
3.

Antigene 3C	Sumitomo
Antioxidant 4010 NA	Bayer
Antiozonant IP	Anchor
Eastozone 34	Eastman a
Flexzone 3C	Uniroyal
Nocrac 810–NA	Ouchi Shinko
Nonox ZA	ICI
Permanax 115	Rhône Poulenc
RC Granulat IPPD	Rhein Chemie
(80 % IPPD, 20 % saturated hydrocarbons and dispersing agents)	
Santoflex IP	Monsanto

4. Brown flakes. M.W. 226.3; M.P. 74 °C. min.; S.G. 1.01–1.07.
   a. Liquid. S.G. 0.90; Flash Point 203 °C.
   Staining oxi, ozo, heat, flex, inh for NR, SBR, NBR, BR, CR.

58
1. N-Methyl-2-pentyl-N'-phenyl-p-phenylenediamine
2.

C_5H_{11}

CH₃—NH—⟨ ⟩—NH—⟨ ⟩

3. Wingstay 300   Goodyear
4. Red-black solid. M.W. 268; M.P. 40–50 °C.; S.G. 0.98.

59
1. N-Phenyl-N'-cyclohexyl-p-phenylenediamine
2. $C_6H_5.NH.C_6H_4.NH.C_6H_{11}$
3. Antioxidant 4010   Bayer

Antiozonant CP   Anchor
  (U.K. only)
Flexzone 6H   Uniroyal
UOP 36   UOP

4. Greyish-white powder. M.W. 267; M.P. 115 °C. min.; S.G. 1.19.
Oxi, ozo, heat, flex, inh for NR, SBR, CR.

60
1. N-Phenyl-N'-hexyl-p-phenylenediamine
2. $C_6H_5.NH.C_6H_4.NH.C_6H_{13}$
3. ANTO$_3$"E"   Pennwalt
4. Greyish solid. M.W. 268; M.P. 45–50 °C.; S.G. 1.015.
Ozo, flex for NR, SBR.

61
1. N-Phenyl-N'-2-octyl-p-phenylenediamine
2. $C_6H_5.NH.C_6H_4.NH.C_8H_{17}$
3. ANTO$_3$"F"   Pennwalt
   UOP 688   UOP
4. Brown viscous liquid. M.W. 296.4. M.P. 10 °C.; B.P. 430 °C.;
   S.G. 1.003.
Ozo, flex for NR, SBR. Staining.

62
1. N-Phenyl-N'-(p-toluene-sulphonyl)-p-phenylenediamine
2.

$CH_3-\langle\rangle-SO_2-NH-\langle\rangle-NH-\langle\rangle$

3. Aranox   Uniroyal
4. Gray powder. M.W. 338; M.P. 140 °C.; S.G. 1.35.
Oxi, inh for NR, CR.

63
1. Diaryl- and alkyl-aryl-p-phenylenediamine blends
2. —
3. Wingstay 250   Goodyear a
   Wingstay 275   Goodyear b
4. a. Dark-brown liquid. Liquid at 38 °C.; S.G. 1.06.
   b. Dark-brown liquid containing crystals. Slushy liquid at room
      temperature. S.G. 1.03.

64
1. N,N'-Bis-(1-methyl heptyl)-p-phenylenediamine / N-(1,3-dimethyl
butyl)-N'-phenyl-p-phenylenediamine / N-phenyl-(N'-methyl
heptyl)-p-phenylenediamine blend
2. —

3. UOP 256   UOP
4. Viscous reddish-brown liquid. Flash-Point 132 °C.; S.G. 0.975.
   Ozo, flex for natural and synthetic rubber.

65 1. N,N'-Diheptyl-p-phenylenediamine / phenyl hexyl-p-phenylenediamine blend
2. —
3. ANTO$_3$"A"   Pennwalt
4. Reddish-brown liquid. S.G. 0.964.
   Staining ozo for NR, SBR.

66 1. N,N'-Bis-(1-methyl heptyl)-p-phenylenediamine / N-phenyl-(N'-
   methyl heptyl)-p-phenylenediamine blend
2. —
3. UOP 62   UOP
4. Viscous reddish-brown liquid. Flash Point 132 °C.; S.G. 0.952.
   Ozo, flex for natural and synthetic rubber.

67 1. N,N'-Bis-(1,4-dimethyl pentyl)-p-phenylenediamine / N-(1,3-di-
   methyl butyl)-N'-phenyl-p-phenylenediamine blend
2. —
3. UOP 57   UOP
4. Viscous reddish-brown liquid. Flash Point 160 °C.; S.G. 0.9638.
   Ozo, flex for natural and synthetic rubber.

68 1. Dioctyl-p-phenylenediamine / phenyl hexyl-p-phenylenediamine
   / phenyl octyl-p-phenylenediamine blend
2. —
3. ANTO$_3$"C"   Pennwalt
4. Reddish-brown liquid. S.G. 0.975.
   Non-staining ozo for natural and synthetic rubber.

Derivatives of diphenylamine

69 1. Alkylated diphenylamine
2. —
3. Pennox A–"S" Powder   Pennwalt a
   Wytox ADP   Nat. Polychemicals b
   Wytox ADP–X   Nat. Polychemicals b

4. a. Grey powder. S.G. 1.26.
   b. Light-tan powder. M.R. 80–88 °C.; S.G. 0.99.
   Oxi for NR.

70 1. p-Isopropoxy-diphenylamine
2.

3. Agerite ISO　　　　　　　　　　　Vanderbilt
4. Tan-to-grey flakes. M.W. 227; M.R. 80–86 °C.; S.G. 1.15.
   Oxi.

71 1. Diphenylamine-derivatives
2. —
3. Antioxidant DDA　　　　　　　　　Bayer
   Antioxidant DDA–EM 30 %　　　　Bayer a
   Antioxidant DDA–EM 50 %　　　　Bayer a
4. Viscous brown to reddish liquid. B.P. 300 °C. min.; S.G. 1.08.
   a. Emulsions for latex.
   Slightly staining oxi, flex, heat, inh for NR, IR, SBR, BR.

72 1. Diphenylamine - acetone condensation product
2. —
3. Accinox B　　　　　　　　　　　　Alkali
   Accinox BL　　　　　　　　　　　Alkali
   Accinox BLN　　　　　　　　　　Alkali
   Agerite Superflex　　　　　　　　Vanderbilt
   Agerite Superflex solid　　　　　Vanderbilt
   Aminox　　　　　　　　　　　　　Naugatuck SpA
   Aminox　　　　　　　　　　　　　Uniroyal
   Anoxin　　　　　　　　　　　　　ACNA Montecatini
   Antigene AM　　　　　　　　　　Sumitomo a
   Antioxidant B　　　　　　　　　　Anchor
   BLE 25　　　　　　　　　　　　　Naugatuck SpA
   BLE 25　　　　　　　　　　　　　Rubber Regenerating
   BLE 25　　　　　　　　　　　　　Uniroyal
   BLE 25 GP　　　　　　　　　　　Naugatuck SpA
   　　(BLE 25 70 %, diatomite 25 %)

BLE 50	Uniroyal
(BLE 50 %, carbon black 50 %)	
Cyanaflex 50	Cyanamid
(absorbed on carbon black)	
Cyanaflex 100	Cyanamid
Neozone L	Du Pont
Nocrac B	Ouchi Shinko
Nonox B	ICI
Nonox B	ICI (India)
Nonox BL	ICI
Nonox BLB	ICI
(absorbed on carbon black)	
Nonox BLN	ICI
Nonox BLW	ICI
(absorbed on silica carrier)	
Permanax 47	Rhône Poulenc

4. Brown powder, liquid, pellets and rods. M.P. 75–95 °C.; S.G. liquid 1.09; S.G. powder 1.14.
a. M.P. 190 °C. min.
Staining ozo, heat, flex for NR, SBR, NBR, CR.

73 1. Diphenylamine - diisobutylene reaction product
2. —
3.

Arrconox DNL	Rubber Regenerating a
Octamine	Rubber Regenerating
Octamine	Naugatuck SpA
Octamine	Uniroyal

4. Light-brown granular waxy solid. M.R. 77–85 °C.; S.G. 0.99.
a. Dark liquid. S.G. 0.95.
Staining oxi, heat, flex for NR, SBR, CR, NBR.

74 1. Diphenylamine - olefin reaction product
2. —
3. Anox NS   Bozzetto
4. Light-brown granular waxy solid. M.R. 75–88 °C.; S.G. 0.98.
Oxi, heat, flex for NR, SBR, CR, IIR.

75 1. Nonylated diphenylamine
2. —

3. Pennox A   Pennwalt
   Polylite   Uniroyal
4. Dark liquid. S.G. 0.95.
   Oxi for NR, SBR, NBR, IR, CR, BR.

76 1. Octylated diphenylamine
   2. —
   3. Agerite Stalite   Vanderbilt
      Agerite Stalite S   Vanderbilt
      Anox NSL   Bozzetto
      Antioxidant OCD   Bayer a
      Antox N   Du Pont
      Cyanox 8   Cyanamid
        (flakes and granules)
      Nonox OD   ICI
   4. a. Gray-brown granules. M.P. 88 °C. min.; S.G. 1.00.
         Oxi, flex, heat, inh for NR, CR.
      Oxi, heat for NR, CR.

77 1. Alkylated diphenylamines and petroleum wax blend (75 : 25)
   2. —
   3. Agerite Gel   Vanderbilt
   4. Soft, light-tan to brown waxy solid. M.R. 40–50 °C.; S.G. 0.94.
      Oxi.

Mixtures and derivatives of alpha- and beta-naphthylamine

78 1. Aldol-alpha-naphthylamine
   2. $N = CHCH_2CH(OH)CH_3$

   3. Aceto AN   Aceto
      Agerite Resin   Vanderbilt
      Antioxidant AP   Bayer
      Antivecchiante AL   ACNA Montecatini
      Nocrac C   Ouchi Shinko
   4. Yellow-brown powder. M.W. 213.28; M.P. 145 °C. min.; S.G. 1.16.
      Staining oxi, ozo for NR, IR, BR, SBR, NBR.

Antidegr.

79 1. Phenyl-alpha-naphthylamine
2.

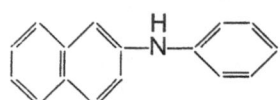

3. Aceto PAN   Aceto
   Antigene PA   Sumitomo
   Antioxidant PAN   Bayer
   Inibitore OA   ACNA Montecatini a
   Neozone A   Du Pont
   Nocrac PA   Ouchi Shinko
   Nonox AN   ICI
4. Light yellowish brown to violet lump. M.W. 219; M.P. 50 °C.
   min.; S.G. 1.21.
   a. S.G. 1.16.
   Staining oxi, flex, heat for NR, SBR, NBR, CR.

80 1. N,N'-Diaryl-p-phenylenediamine / phenyl-alpha-naphthylamine
   blend
2. —
3. Antigene FC   Sumitomo
4. Dark-brown lump.
   Staining oxi, heat, flex for NR, SBR, CR.

81 1. Phenyl-alpha-naphthylamine / 2,4-toluene diamine blend
   (92.5 : 7.5)
2. —
3. Neozone C   Du Pont
4. Gray-brown pellets. M.P. 46 °C.; S.G. 1.22.
   Staining oxi, heat for NR, SBR, IIR.

82 1. Phenyl-beta-naphthylamine
2.

3. Accinox D   Alkali
   Aceto PBN   Aceto

Agerite Powder	Vanderbilt
Antigene D	Sumitomo
Antigene DF	Sumitomo
Antioxidant 116	Anchor
Antioxidant PBN	Bayer
Antioxidant PBN	Spolek
Inibitore OD	ACNA Montecatini
Neozone D	Du Pont
Neozone D Special	Du Pont
(easy dispersable, for use	
in latex)	
Nocrac D	Ouchi Shinko
Nonox D	ICI
Nonox D	ICI (India)
Nonox DN	ICI
(powder and flakes)	
PBN	BASF
PBN	Monsanto
PBN	Uniroyal

4. Grayish-white powder. M.W. 219; M.P. 105 °C. min.; S.G. 1.23.
Staining oxi, flex for NR, SBR, NBR, CR.

83 1. N,N'-Diaryl-p-phenylenediamine / phenyl-beta-naphthylamine
blend

2. —

3. Akroflex DAZ (1 : 1)   Du Pont a
Antigene HP–S   Sumitomo b

4. a. Dark-grey powder. M.P. 70 °C. min.; S.G. 1.27.
Oxi, ozo for CR.

b. Light-purple powder. M.P. 90 °C. min.
Oxi, ozo, heat, flex for NR, SBR, NBR, CR.

84 1. N,N'-Diphenyl-p-phenylenediamine / phenyl-beta-naphthylamine
blend

2. —

3. Agerite HP (65 : 35)   Vanderbilt
Agerite HPX   Anchor
(Antioxidant 108X outside U.K.)

Akroflex CD (65 : 35)   Du Pont
Antigene HP   Sumitomo
Nocrac HP   Ouchi Shinko
Nonox HP   ICIANZ
4. Grey to light-purple powder, rods, pellets. M.R. 89–94 °C.; S.G.
   1.24.
   Staining oxi, flex for NR, SBR, NBR, CR.

85 1. N,N'-Diphenyl-p-phenylenediamine / 4,4'-dimethoxy-diphenyl-
       amine / N-phenyl-beta-naphthylamine blend (25 : 25 : 50)
    2. —
    3. Thermoflex A   Du Pont
    4. Gray-to-black pellets. M.P. 78 °C.; S.G. 1.22.
       Oxi, ozo, flex, heat for NR, SBR, CR.

86 1. N,N'-Diphenyl-p-phenylenediamine / p-isopropoxy-diphenyl-
       amine / phenyl-beta-naphthylamine blend (25 : 25 : 50)
    2. —
    3. Agerite Hipar   Vanderbilt
    4. Gray-to-brown powder and rods. M.R. 60–75 °C.; S.G. 1.16.
       Oxi.

87 1. Phenyl-beta-naphthylamine - acetone reaction product
    2. —
    3. Antigene DA   Sumitomo a
       Betanox Special   Uniroyal b
    4. a. Light-brown powder. M.P. 65 °C. min.
       b. Tan powder. M.P. 130 °C. min.; S.G. 1.14.
       Oxi, heat, flex for NR, SBR, NBR.

88 1. N,N'-Diaryl-p-phenylenediamine / phenyl-beta-naphthylamine -
       acetone reaction product blend
    2. —
    3. Antigene FL   Sumitomo
    4. Dark-gray powder.
       Heat, flex for NR, SBR, CR.

Heterocyclic compounds

89 1. Benzofurane derivative
2. —
3. Antioxidant AFC   Bayer
4. Light-brown powder. M.P. 160 °C.; S.G. 1.25.
   Oxi, flex, heat for solid rubber and latex. Ozo for CR and for
   CR-mixtures.

90 1. 2-(3,5-Di-tert butyl-4-hydroxyanilino)-4,6-bis-(N-octylthio)-1,3,5-
   triazine
2.

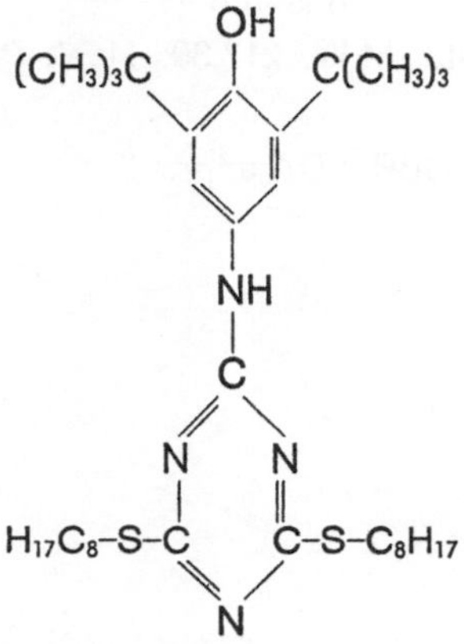

$(CH_3)_3C$ — OH — $C(CH_3)_3$
NH
C
N   N
$H_{17}C_8$–S–C   C–S–C_8H_{17}
N

3. Irganox 565   Ciba Geigy
4. White crystalline powder. M.W. 589; M.P. 91–96 °C.
   Oxi.

91 1. 6-Dodecyl-1,2-dihydro-2,2,4-trimethyl-quinoline
2.

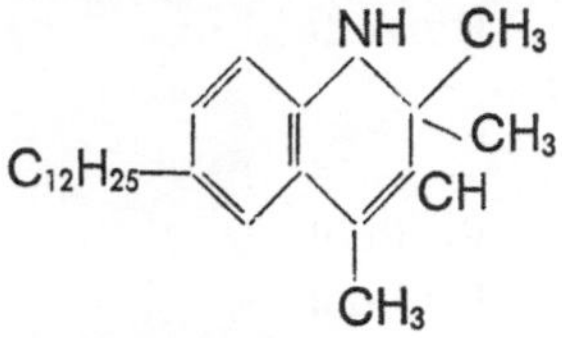

$C_{12}H_{25}$ — NH CH_3
CH_3
CH
CH_3

3. Santoflex DD   Monsanto
4. Dark viscous liquid. M.W. 341.56; S.G. 0.90–0.96.
   Staining oxi, flex for NR. FDA appr. Stabilizer for SBR during
   storage and production.

92 1. 6-Ethoxy-2,2,4-trimethyl-1,2-dihydroquinoline

2.

3.
Anox W	Bozzetto
Antioxidant EC	Bayer
Ethoxyquin	Bennett
Ethoxyquin	Raschig
Nocrac AW	Ouchi Shinko
Santoflex AW	Monsanto

4. Dark red-brown viscous liquid. M.W. 217.30; Flash Point 162–164 °C.; S.G. 1.02–1.05.

Staining oxi, ozo, flex for NR, SBR. FDA appr.

93 1. Ethylene-thiourea
syn. 2-Mercaptoimidazoline

2.

3.
Accelerator MI 12	Metallgesellschaft
—	Degussa
E.T.U.	Hasselt
E.T.U.	Prochim
JOR 4022	Bozzetto
JOR 4022 oleato	Bozzetto
(83 % JOR 4022, 17 % paraffin oil)	
NA–22	Du Pont
NA–22 D	Du Pont
(80 % dispersion of NA–22 in oil)	
Nocceler 22	Ouchi Shinko
Pennac CRA	Pennwalt

Pennac CRA   Vondelingenplaat
Robac 2.2   Robinson
Rodanin S 62   Dimitrova
Sanceler 22   Sanshin
Soxinol 22   Sumitomo
Vulkacit NPV/C   Bayer
(coated)

4. White powder. M.W. 102.16; M.P. 195 °C. min.; S.G. 1.43–1.45.
Ozo for NR; non-staining acc. for CR.

△ Acc. 27

94 1. 2-Mercaptobenzimidazole

2.

3. Antigene MB   Sumitomo
Antioxidant MB   Dimitrova
Antioxidant MB   Bayer
Antivecchiante MB   ACNA Montecatini
MBI   Prochim
Nocrac MB   Ouchi Shinko
Permanax 21   Rhône Poulenc
Vondantox MBI   Vondelingenplaat

4. Yellow-white powder. M.W. 150; M.P. 290 °C. (with decomposition); S.G. 1.42.
Non-staining oxi, heat, inh for NR, SBR, NBR; acc. for NR; pept. for CR.

△ Acc. 86
△ Pept. 3

95 1. Poly-2,2,4-trimethyl-1,2-dihydroquinoline

2.

3. Aceto POD   Aceto
   Agerite AK   Anchor
      (tradename outside U.K. Anti-
      oxidant 184)
   Agerite Resin D   Vanderbilt
   Anox HB   Bozzetto
   Antigene RD   Sumitomo
   Antioxidant HS   Bayer a
   Flectol H   Monsanto
   Nocrac 224   Ouchi Shinko
   Pennox HR   Pennwalt
   Permanax 45   Rhône Poulenc
4. Buff powder or flakes. M.P. 100–120 °C.; S.G. 1.08.
   a. M.P. 75 °C. min.
   Semi-staining oxi, heat, inh for NR, NBR, BR, IR, CR, also for
   latex; vulc. for CR.
   △ Vulc. 50

96 1. Tetraethyl-thiuram disulphide
    2. $(C_2H_5)_2N.CS.S.S.CS.N(C_2H_5)_2$
    3. Accicure TET   Alkali
        Aceto TETD   Aceto
        Ancazide ET   Anchor
        Ethyl Thiram   Pennwalt
        Ethyl Thiurad   Monsanto
        Ethyl Tuads   Vanderbilt a
        Ethyl Tuex   Naugatuck SpA
        Ethyl Tuex   Uniroyal
        Etiurac   Ticino
        Eveite T   ACNA Montecatini
        Hermat TET   Dimitrova
        Nocceler TET   Ouchi Shinko
        Robac TET   Robinson
        Sanceler TET   Sanshin
        Soxinol TET   Sumitomo
        Superaccelerator 481   Rhône Poulenc
        TETD   Hasselt
        TETD   Prochim

TETS   Bozzetto
Thiuram E   Du Pont
Vondac TET   Vondelingenplaat
Vulcafor TET   ICI
Vulcafor TET   ICI (India)

4. Grayish-white powders and pellets. M.W. 296.54; M.P. 71–73 °C.;
   S.G. 1.26.
   a. Powder and rods.
   Non-staining stabilizer for Neoprene G; acc. for NR, SBR, NBR,
   IIR, IR, EPDM; act. for thiazole-, guanidine- and aldehyde-
   type acc.; vulc. for sulphurless compounds.
   △ Acc. 81
   △ Act. 8
   △ Vulc. 36

97 1. Trimethyldihydroquinoline-derivative
     2. —
     3. Antigene MW   Sumitomo
     4. Dark-brown viscous liquid.
        Staining oxi, heat, flex for synthetic dry rubber and latex.

98 1. Zinc 2-mercaptobenzimidazole
     2.

     3. Antigene MBZ   Sumitomo
        Antioxidant ZMB   Bayer
        M.B.I. Zn   Prochim
        Nocrac MBZ   Ouchi Shinko
        Permanax Z21   Rhône Poulenc
     4. White powder. M.W. 363.4; M.P. 300 °C. (with decomposition);
        S.G. 1.6.
        Non-staining oxi, heat, inh for NR, SBR, NBR.

99 1. Zinc 2-mercaptobenzothiazole

2.

3.

Accicure ZMBT	Alkali
Bantex	Monsanto
Eveite MZ	ACNA Montecatini
Hermat Zn MBT	Dimitrova
Nocceler MZ	Ouchi Shinko
OXAF	Naugatuck SpA
OXAF	Uniroyal
Pennac ZT	Pennwalt
Pennac ZT-"W"	Pennwalt
(containing 10 % inert hydro-	
carbon)	
Rapid Accelerator 205	Rhône Poulenc
Sanceler MZ	Sanshin
Soxinol MZ	Sumitomo
Vulcafor ZMBT	ICI
Vulcafor ZMBT	ICI (India)
Vulkacit ZM	Bayer
Zenite	Du Pont
Zenite Special	Du Pont
Zetax	Vanderbilt
Zinc Ancap	Anchor
ZMBT	Cyanamid
ZMBT waxed	Cyanamid
ZMBT wettable	Cyanamid

4. Light cream powder. M.W. 397.9; M.P. 300 °C. (with decomposition); S.G. 1.72.
   Non-staining oxi for latex; acc. for NR, SBR, NBR latices.
   △ Acc. 100

Dithiocarbamates

---

100 1. Dibutylammonium-dibutyldithiocarbamate
2. $(C_4H_9)_2.N.CS.S.NH_2(C_4H_9)_2$
3. Robac D.B.U.D.   Robinson

4. Brown flakes. M.W. 334.6; M.P. 45 °C.
   Oxi for rubber-based adhesives; non-staining acc. for NR- and
   SBR proofing compounds.
   △ Acc. 46

101 1. Nickel dibutyldithiocarbamate
2. [(C$_4$H$_9$)$_2$N.CS.S–]$_2$Ni
3.

Antigene NBC	Sumitomo
JO 4011	Bozzetto
NBC	Du Pont
NDBC	Hasselt
Nilame	Prochim
Nocrac NBC	Ouchi Shinko

4. Green powder. M.W. 467.47; M.R. 86–90 °C.; S.G. 1.10.
   Oxi, ozo, heat, flex for SBR, CR, CSM.

102 1. 2,2'-Dibenzothiazyl disulphide - morpholinium-N-oxy-diethylene-
   dithiocarbamate reacton product
2. —
3. Antigene 3M   Sumitomo
4. White powder. M.P. 105 °C.
   Non-staining oxi, ozo, flex for NR, SBR.

103 1. Zinc dibutyldithiocarbamate
2. [(C$_4$H$_9$)$_2$N.CS.S–]$_2$Zn
3.

Aceto ZDBD	Aceto
Ancazate BU	Anchor
Butazate	Naugatuck SpA
Butazate	Uniroyal
Butazate 50D	Uniroyal
(50 % slurry for use in latex)	
Butazin	Ticino
Butyl Zimate	Vanderbilt
Butyl Ziram	Pennwalt
Butyl Ziram	Pennwalt
(50 % aquous solution)	
Butyl Zirame	Prochim
Eptac 4	Du Pont
Eveite Butil Z	ACNA Montecatini
JO 4013	Bozzetto

Nocceler BZ	Ouchi Shinko
Robac Z.B.U.D.	Robinson
Sanceler BZ	Sanshin
Soxinol BZ	Sumitomo
Ultra Accelerator DI 13	Metallgesellschaft
Vondac ZBUD	Vondelingenplaat
Vulcafor ZNBC	ICI †
ZDBC	Hasselt

4. Cream powder. M.W. 474.14; M.R. 95–108 °C.; S.G. 1.21.
   Antidegr. for unvulcanized rubber and for non-staining grades of IIR; non-staining ultra-acc. for NR, BR, SBR, NBR and their latices, acc. for EPDM; act. for thiazole- and other acid-type acc.
   △ Acc. 65
   △ Act. 28

Miscellaneous, mixtures and undisclosed compositions

104 1. Alkylated aryl-phosphites blend
2. —
3. Alkanox   Bozzetto
   Antioxidant 6   Dimitrova
4. Amber liquid.
   Non-staining oxi for NR, SBR, BR.

105 1. 2,5-Di-tert amyl hydroquinone
2.

$$CH_3CH_2-\overset{\overset{\displaystyle CH_3}{|}}{\underset{\underset{\displaystyle CH_3}{|}}{C}}\!\!-\!\!\!\left\langle\!\!\!\begin{array}{c}OH\\ \\ \\OH\end{array}\!\!\!\right\rangle\!\!-\!\!\overset{\overset{\displaystyle CH_3}{|}}{\underset{\underset{\displaystyle CH_3}{|}}{C}}\!\!-\!\!CH_2CH_3$$

3. Antioxidant DITAH   Conestoga
   Santovar A   Monsanto
   —   Eastman
4. Off-white powder. M.W. 250.37; M.P. 176 °C. min.; S.G. 1.02–1.08.
   Stabilizer for NBR; non-staining oxi for uncured rubber. FDA appr.

106 1. Diarylamine - ketone - aldehyde reaction product
2. —
3. BXA   Naugatuck SpA
   BXA   Uniroyal
4. Brown powder. M.R. 85–95 °C.; S.G. 1.10.
   Staining oxi, ozo, heat, flex for NR, SBR, NBR, CR.

107 1. complex Diarylamine - ketone reaction product / N,N'-diphenyl-
   p-phenylenediamine blend (65 : 35)
2. —
3. Flexamine   Uniroyal
   Flexamine G   Uniroyal
4. Brown powder and granules. M.R. 75–90 °C.; S.G. 1.20.
   Oxi, ozo, heat, flex, inh for NR, SBR, IR, BR.

108 1. 2,5-Di-tert butylquinol
   syn. 2,5-Di-tert butyl-hydroquinone
2.

3. Nocrac NS–7   Ouchi Shinko
   —   Eastman
   —   May & Baker
4. Brownish-grey powder. M.W. 222.33; M.P. 200 °C.
   Oxi for NR, IR, BR, SBR, NBR, CR.

109 1. N,N'-Dibutylthiourea
2. $(C_4H_9NH)_2CS$
3. Accelerator DBT   BASF
   DBTU   Prochim
   —   Degussa
   Pennzone B   Pennwalt
   Pennzone B   Vondelingenplaat
   Robac DBTU   Robinson

4. Off-white powder. M.W. 188.3; M.P. 65 °C.
Antidegr. for NR-latex and for thermoplastic SBR; acc. for mercaptan-modified CR; act. for EPDM and NR.
△ Acc. 22
△ Act. 35

110 1. 1,3-Diethylthiourea
2. $(C_2H_5NH)_2CS.$
3.

DETU	Prochim
JOR 4050	Bozzetto
—	Degussa
Pennzone E	Pennwalt a
Pennzone E	Vondelingenplaat a
RC Granulat DETU	Rhein Chemie
(80 % DETU, 20 % saturated hydrocarbons and special dispersion agents)	
Robac DETU	Robinson

4. Yellow powder. M.W. 132.2; M.P. 75 °C.; S.G. 0.98. Water-soluble.
a. White flakes. S.G. 1.12.
Antidegr. for NR, NBR, SBR, CR; acc. for mercaptan-modified CR (Neoprene W).
△ Acc. 23

111 1. Dilaurylthiodipropionate
2. $(CH_2CH_2CO_2C_{12}H_{25})_2S$
3.

Antigene TPL	Sumitomo
DLTDP	Reagens
Irganox PS 800	Ciba Geigy

4. White crystalline powder or flakes. M.W. 514; M.P. 38 °C. min.
Oxi, ozo, heat, flex for NR, SBR.

112 1. p-Dimethoxybenzene
syn. Quinol-dimethylether
Hydroquinone-dimethylether

2. 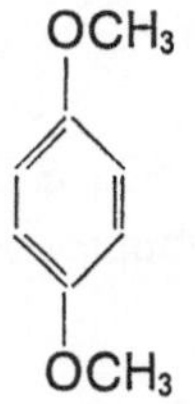   OCH₃

    ÓCH₃

3. —   May & Baker
4. White crystalline solid. M.W. 138.2; M.P. 56 °C.; S.G. 1.053.
   Oxi.

113 1. N,N'-Diphenyl-p-phenylenediamine / acrylamide-derivative
        blend
     2. —
     3. Accinox DPL   Alkali
     4. Dark brown viscous liquid. S.G. 1.16.
        Oxi.

114 1. N,N'-Diphenyl-p-phenylenediamine / 6-dodecyl-1,2-dihydro-2,2,4-
        trimethylquinoline blend
     2. —
     3. Santoflex 75   Monsanto
     4. Dark flakes. S.G. 1.09–1.15.
        Oxi.

115 1. Dioctadecylthiodipropionate
     2. (CH₂CH₂CO₂C₁₈H₃₇)₂S
     3. Irganox PS 802   Ciba Geigy
     4. White crystals. M.W. 682.0; M.P. 63–65 °C.
        Oxi, co-stabilizer.

116 1. Di-ortho-tolylguanidine salt of dicatecholborate
     2. 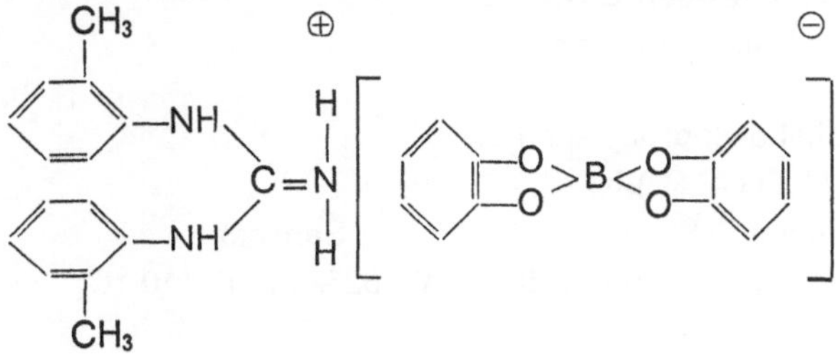   CH₃   ⊕   ⊖

3. Nocceler PR   Ouchi Shinko
   Permalux   Du Pont
4. Grayish-brown powder. M.W. 482.8; M.P. 165 °C.; S.G. 1.25.
   Antidegr. for NR, SBR; non-staining acc. for Neoprene-G types.
   △ Acc. 33

117 1. Hydroquinone monobenzylether
2. —CH$_2$-O—OH
3. Agerite Alba   Vanderbilt
4. Light-tan powder. M.W. 200; M.R. 108–115 °C.; S.G. 1.26.
   Oxi.

118 1. Hydroquinone-monomethylether
   syn. p-Methoxy phenol
2. $CH_3OC_6H_4OH$
3. MEHQ   Conestoga
4. Flakes. M.W. 124.13; M.P. 53 °C.
   Oxi.

119 1. 2-Mercaptobenzimidazole / sym. di-beta-naphthyl-p-phenylene-
   diamine blend
2. —
3. Nonox CGP   ICI †
4. Grey powder. S.G. 1.34.
   Oxi, inh.

120 1. 2-Mercaptobenzimidazole / 2,2'-methylene-bis-[6-(alpha-methyl-
   cyclohexyl)-p-cresol] blend
2. —
3. Nonox CNS   ICI
4. Cream powder. S.G. 1.25.
   Non-staining oxi, inh.

121 1. Nickel diisopropylxanthate
2. $(C_3H_7.O.CS.S-)_2Ni$
3. Sandant PN   Sanshin
4. Yellowish-green powder. M.W. 325.4; M.P. 110 °C.
   Flex.

122 1. Thiourea-derivatives
     2. —
     3. Nocrac NS–10–N   Ouchi Shinko
        Nocrac NS–11   Ouchi Shinko
     4. Oxi for NR, IR, SBR, NBR, CR.

123 1. Tributylthiourea
     2. $(C_4H_9)_2N.CS.NHC_4H_9$
     3. Santowhite TBTU   Monsanto
     4. Pale-amber liquid. M.W. 244; S.G. 0.938.
        Non-staining ozo for natural and synthetic rubber.

124 1. Tris-(nonylated phenyl) phosphite
     2.

$$\left[C_9H_{19}\!-\!\!\!\diagbox\!\!\!-O \right]_3\!\!-P$$

     3. Antigene TNP   Sumitomo
        Anullex TNPP   Pearson
        Irgafos TNPP   Ciba Geigy
        Naugard   Uniroyal
          (in Europe Polygard)
          Grades P, P powder (68 %
          active ingredient on calcium-
          silicate carrier), PHR (hydro-
          lysis resistant)
        Nocrac TNP   Ouchi Shinko
        Polygard   Rubber Regenerating
        Polygard   Naugatuck SpA
     4. Light-yellow tacky liquid. M.W. 689; S.G. 0.99.
        Non-staining oxi, heat, flex for NR, SBR, NBR.

125 1. Composition undisclosed
     2. —
     3. Anox SD   Bozzetto
        Antioxidans GI–08–288   Ciba Geigy
        Antioxidant 439   Uniroyal
        Antioxidant 451   Uniroyal
        Antioxidant TD EM 50   Bayer
        Antivecchiante ADD   ACNA Montecatini

Antidegr.

Arrconox DNP	Rubber Regenerating
Arrconox GP	Rubber Regenerating
Hallcolite OP	Hall
Inibitore AT	ACNA Montecatini
Kenmix	Kenrich a
Nevastain A	Neville
Nevastain B	Neville
Nonox	ICI
Grades CC, EX, EXN, EXP, NS, WSL, WSO	
Oxystop 999	Arwal
Ozonschutzmittel AFD	Bayer
Ozonstop	Arwal
Grades 6000–E, 6000–P	
Raluquin K	Raschig
Wytox 345	Nat. Polychemicals

4. a. △ Acc. 141
   △ Act. 44
   △ Vulc. 54

Blowing agents

INORGANIC

1 1. Sodiumbicarbonate
2. NaHCO₃

   2. $NaHCO_3$

3. —   . Anchor a
   Isocell S   Rhein Chemie b
   Sponge Paste 2   Ouchi Shinko c
   Sponge Paste 3   Ouchi Shinko d
   Treibmittel 1843   Rhein Chemie e
   Unicel S   Du Pont f
   Unicel SX   Du Pont g

4. a. White powder. M.W. 84.02; M.P. 270 °C. (with decomposition); S.G. 2.15.
   b. Grayish coarse powder. Decomposition at 70 °C.; S.G. 2.15.
   c. 80 % $NaHCO_3$.
   d. 70 % $NaHCO_3$.
   e. Grayish coarse powder. Decomposition at 70 °C.; S.G. 2.15.
   f. 50 % Dispersion in mineral oil. Decomposition at 100 °C.; S.G. 1.30.
   g. 70 % Dispersion in mineral oil. Decomposition at 100 °C.; S.G. 1.55.

ORGANIC

Azo-compounds

2 1. Azo-dicarbonamide
       syn. 1,1-Azo-bis-formamide
2. $NH_2CON:NCONH_2$
3. Alveofer AZDC   Bozzetto
   Azocel   Fairmount
      Grades 504, 508, 525, varying
      in particle size.
   BZ M2   Organo Synthèse
   Celogen AZ   Uniroyal

Dispercel   Fairmount
(50 % dispersion of Azocel
in DOP)

Genitron AC   Fisons
Grades AC/2, AC/3, AC/4
varying in particle size, also
available as paste in plastici-
zer. (Trade name in N- and S-
America and in Japan Ficel.)

Isocell AC   Rhein Chemie †
Kempore   Nat. Polychemicals
Noury ADC   Noury
Grades ADC/4, ADC/8,
ADC/25 varying in color
and in particle size.

Porofor ADC/R   Bayer

4. Yellow powder. M.W. 116.08; S.G. 1.63; decomposition temperature in air 195–200 °C., in polymers 150–200 °C.

3 1. Azo-diisobutyronitrile
2. $(CH_3)_2.(CN).C-N=N-C(CN).(CH_3)_2$
3. Genitron AZDN   Fisons
(trade name in N- and S-
America and in Japan Ficel)
Porofor N   Bayer
4. White powder. M.W. 164.2; S.G. 1.11; decomposition temperature 103 °C.

4 1. Diazoaminobenzene
2.

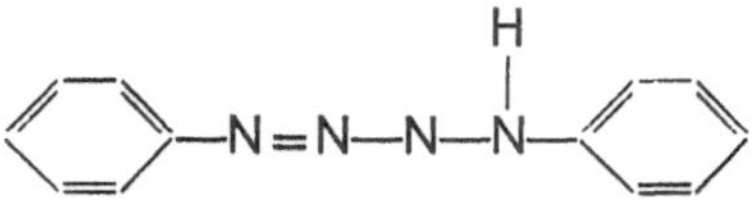

3. Vulcacel AN   ICI †
4. Brown powder. M.W. 197.

Nitroso-compounds

5 1. N,N'-Dimethyl-N,N'-dinitrosoterephthalamide

2.

H₃C-N-C- ... -C-N-CH₃ with O, NO groups

3. Nitrosan   Du Pont
4. Yellow powder. M.W. 250; decomposition range 80–100 °C.;
   S.G. 1.20.
   Non-staining.

6   1. Dinitrosopentamethylenetetramine
        2. $H_2C — N — CH_2$

             $ON — N \quad CH_2 \quad N — NO$

             $H_2C — N — CH_2$

        3. Alveofer DNP 80   Bozzetto
             (80 % DNP, 20 % inert fillers)
           Alveofer DNP/GM   Bozzetto
             (80 % DNP, 20 % inert fillers
             and dispersion agents)
           Chempor   Chemko
             Grades B 80, N 90, PC 50,
             PC 65, PC 80.
           DNPT   Organo Synthèse
             (74 % active ingredient)
           Esporen   ACNA Montecatini
           Naftopor   Metallgesellschaft
           Nocblow DPT   Ouchi Shinko
           Opex   Nat. Polychemicals
             Grades 40, 42, 80, 93, 100
             (% DNPT)
           Porofor DNO/F   Bayer
             (80 % DNPT, 20 % inert fillers)
           Porofor DNO/N   Bayer
             (80 % DNPT, 20 % inert fillers)
           Sponge Paste 1   Ouchi Shinko
             (50 % active ingredient)
           Sponge Paste 4   Ouchi Shinko
             (50 % active ingredient)

Unicel	Du Pont
Grades 80, 93, 100 (% active ingredient) and ND (1 part Unicel 100 = 2.5 paste Unicel ND)	
Vondablow	Vondelingenplaat
(80 % active ingredient)	
Vulcacel BN 94	ICI

4. Pale-yellow powder. M.W. 186.17; decomposition temperature in air 190–200 °C.; S.G. 1.51.

Sulfo(nyl)hydrazide-compounds

7
1. Benzene disulfohydrazide
2. $C_6H_4.(SO_2.NH–NH_2)_2$
3. Porofor B 13/CP 50   Bayer
   (Paste with 50 % chlorinated paraffin)
4. Yellowish paste. M.W. 266; M.P. 147 °C.; S.G. 1.5; decomposition temperature 155 °C.

8
1. Benzene sulfohydrazide
2. $C_6H_5.SO_2.NH.NH_2$
3. BSH   Organo Synthèse
   Celogen BSH   Uniroyal
   Celogen BSH paste   Uniroyal
   (75 % BSH, 25 % paraffin oil)
   Genitron BSH   Fisons
   (trade name in N- and S-America and in Japan Ficel)
   Porofor BSH powder   Bayer
   Porofor BSH paste   Bayer
   (75 % BSH, 25 % paraffin oil)
   Porofor BSH paste M   Bayer
   (75 % BSH, 25 % paraffin oil, micronized)
4. Cream-to-buff powder. M.W. 172.2; decomposition temperature 99 °C.; S.G. 1.43.

9 1. p,p'-Oxy-bis (benzene sulphonyl) hydrazide
     syn. Diphenyloxide-4,4'-disulphonhydrazide
  2. $NH_2.NH.SO_2.C_6H_4.O.C_6H_4.SO_2.NH.NH_2$
  3. Celogen OT   Uniroyal a
     Genitron OB   Fisons b
       (trade name in N- and S-
       America and in Japan Ficel)
     Nitropore OBSH   Nat. Polychemicals c
  4. White crystalline powder. M.W. 358.4
     a. S.G. 1.55; decomposition temperature 158–160 °C.
     b. S.G. 1.43; decomposition temperature 150 °C.
     c. S.G. 1.54; decomposition temperature 127–149 °C.

10 1. p-Toluene sulphohydrazide
  2.

$$CH_3-C_6H_4-\overset{\displaystyle O}{\underset{\displaystyle O}{\overset{\|}{\underset{\|}{S}}}}-\overset{\displaystyle H}{\overset{|}{N}}-NH_2$$

  3. Celogen TSH   Uniroyal
     Isocell TSH   Rhein Chemie †
  4. Cream crystalline powder. M.W. 186; M.R. 100–110 °C.; S.G.
      1.42; decomposition temperature 110–120 °C.
     For NR, SBR, CR and polysulphiderubber.

11 1. Sulphohydrazides
  2. —
  3. Alveofer BSI   Bozzetto
     Alveofer BSI Paste   Bozzetto
       (75 % BSI, 25 % paraffin oil)
  4. White powder. M.P. 100 °C. min.; S.G. 1.4.

Miscellaneous, mixtures and undisclosed compositions

---

12 1. Alkylsulphonate and activating additives blend
  2. —
  3. Levapon TH   Bayer
       (highly concentrated)

4. Yellowish paste. S.G. 1.04.
   Secondary blowing agent in combination with sodiumoleate to produce latex foams. Foam-stabilizer in the manufacture of solid cellular rubber goods.

13 1. Amino-guanidine bicarbonate
2. $[NH_2.NH.C(:NH).NH_2].HCO_3$
3. Genitron OC   Fisons
   (trade name in N- and S- America and in Japan Ficel)
4. Off-white powder. M.W. 136.1. Decomposition temperature 167 °C.

14 1. p-Toluene-sulfonyl-semicarbazide
2. $CH_3.C_6H_4.SO_2.NH.NH.CO.NH_2$
3. Celogen RA   Uniroyal
4. Off-white powder. M.W. 136.1; decomposition temperature 167 °C.

15 1. Zinc-amine complex
2. —
3. Ancablo A   Anchor
4. White powder.
   Ba. for NR, SBR, IIR, CR.

16 1. Composition undisclosed
2. —
3. Genitron CR   Fisons a
   Genitron THT   Fisons b
   (trade name in N- and S- America and in Japan Ficel)
   OP–2   Organe Synthèse c
4. a. Yellow powder. Decomposition temperature 150–200 °C.
   b. Off-white powder. Decomposition temperature 265–290 °C.
   c. Yellow powder. Decomposition temperature 150–160 °C.

Co-agents

---

1   1. 1,3-Butylideneglycoldimethacrylate

2.
$$CH_2=\underset{\underset{O}{\parallel}}{\overset{\overset{CH_3}{|}}{C}}-C-O-CH_2-CH_2-\overset{\overset{CH_3}{|}}{CH}-O-\underset{\underset{O}{\parallel}}{C}-\overset{\overset{CH_3}{|}}{C}=CH_2$$

3. SR 297   Anchor
   SR 297   Sartomer

4. Clear straw coloured liquid. M.W. 226; B.P. 290 °C.; S.G. 1.009.
   Co-agent for the peroxide vulc. of EPDM and NBR.

2   1. Ethylenedimethacrylate
   syn. Ethyleneglycoldimethacrylate

2.
$$CH_2=\underset{\underset{O}{\parallel}}{\overset{\overset{CH_3}{|}}{C}}-C-O-CH_2-CH_2-O-\underset{\underset{O}{\parallel}}{C}-\overset{\overset{CH_3}{|}}{C}=CH_2$$

3. SR 206   Anchor
   SR 206   Sartomer

4. Clear, water-white liquid. M.W. 198; B.P. 260 °C.; S.G. 1.05.
   Co-agent in peroxide vulcanization of EPDM and NBR.

3   1. Phenolformaldehyde resin

2. —

3. Vulkadur A   Bayer

4. White to yellow powder. M.P. 75–90 °C.; S.G. 1.27.
   Reinforcing resin for NBR.

4   1. N,N'-m-Phenylene-dimaleimide

2.
$$\underset{\underset{O}{\parallel}}{\overset{\overset{O}{\parallel}}{\underset{HC-C}{HC-C}}}>N-\langle\text{phenylene}\rangle-N<\underset{\underset{O}{\parallel}}{\overset{\overset{O}{\parallel}}{\underset{C-CH}{C-CH}}}$$

3. HVA 2   Du Pont

4. Yellow powder. M.W. 268; M.P. 200 °C.; S.G. 1.44.
   Co-agent especially for CSM vulcanization.

5 1. Trimethylolpropanetrimethacrylate

2.

$$
\begin{array}{c}
CH_3 \\
| \\
CH_2{=}C{-}C{-}O \\
\| \quad | \\
O \quad | \\
\end{array}
$$

$$
CH_2{=}C{-}C{-}O{-}CH_2{-}C{-}CH_2{-}CH_3
$$

with the methacrylate groups:

- $CH_2{=}\overset{\overset{\textstyle CH_3}{|}}{C}{-}\overset{}{C}{-}O$ (top, with O double bond)
- $CH_2{=}\overset{\overset{\textstyle CH_3}{|}}{C}{-}\overset{}{C}{-}O{-}CH_2{-}\overset{\overset{\textstyle CH_2}{|}}{C}{-}CH_2{-}CH_3$
- $CH_2{=}\overset{\overset{\textstyle CH_3}{|}}{C}{-}\overset{}{C}{-}O$ (bottom, with O double bond)

3. SR 350   Anchor

     SR 350   Sartomer

4. Clear straw-coloured liquid. M.W. 338; B.P. 200 °C. at 1 mm min.

    Co-agent for the peroxide vulcanization of NBR and EPDM.

Peptizers

---

1   1. 4-tert Butyl-o-thiocresol

2.

3. Pitt Consol 646   Pitt Consol
   Pitt Consol 646 powder   Pitt Consol
   (55 % active ingredient on
   SiO$_2$ filler)

4. 55 % Hydrocarbon solution; colourless, low-viscosity liquid.
   Flash-Point 68.3 °C.; S.G. 0.87–0.90.
   Pept. for NR, SBR and other synthetic rubbers.

2   1. o,o'-Dibenzamido-diphenyl disulphide
   syn. Di-thio-bis-benzanilide
       Di-(o-benzamidophenyl) disulphide

2.

3. Noctizer SS   Ouchi Shinko
   Peptazin BAFD   Dimitrova
   Peptisant 10   Rhône Poulenc
   Pepton 22   Cyanamid
   Pepton 22   Anchor

4. Pale-yellow powder. M.W. 456.59; M.P. 136 °C.; S.G. 1.35.
   Pept. for NR, BR, IR, SBR.

3   1. 2-Mercaptobenzimidazole

2.

3. Antigene MB   Sumitomo
   Antioxidant MB   Dimitrova
   Antioxidant MB   Bayer

Antivecchiante MB	ACNA Montecatini
MBI	Prochim
Nocrac MB	Ouchi Shinko
Permanax 21	Rhône Poulenc
Vondantox MBI	Vondelingenplaat

4. Yellow-white powder. M.W. 150; M.P. 290 °C. (with decomposition); S.G. 1.42.

Pept. for CR; acc. for NR; non-staining oxi, heat, inh for NR, SBR, NBR.

△ Acc. 86

△ Antidegr. 94

4 1. 2-Mercaptobenzothiazole

2.

3.

Accicure MBT	Alkali
Ancap	Anchor
Captax	Vanderbilt
Eveite M	ACNA Montecatini
MBT	Akron
MBT	Du Pont
MBT	Naugatuck SpA
MBT	Uniroyal
MBT	Cyanamid
MBT XXX	Cyanamid
(very pure grade)	
Nocceler M	Ouchi Shinko
Pennac MBT	Pennwalt
Pneumax MBT	Dimitrova
Rapid Accelerator 200	Rhône Poulenc
Rotax	Vanderbilt
Sanceler M	Sanshin
Soxinol M	Sumitomo
Thiotax MBT	Monsanto
Vulcafor MBT	ICI
Vulcafor MBT	ICI (India)
Vulkacit Merkapto	Bayer

4. Light yellow powder. M.W. 167.25; M.R. 164–175 °C.; S.G. 1.50.
   Pept. for NR; non-staining semi-ultra acc. for NR, SBR, NBR,
   IIR, also for latex; ret. for CR.
   △ Acc. 96
   △ Ret. 6

5 1. Pentachlorothiophenol (with additives)
   syn. Pentachlorophenyl-mercaptan
   2. C_6Cl_5SH
   3. Ciclizante 2°B   Bozzetto a
      Ciclizante 3°C   Bozzetto b
      Renacit V   Bayer c
      Renacit VII   Bayer d
      RPA No. 6   Du Pont e
   4. Gray powder. M.W. 280; S.G. a. 1.15, b. 1.17, c. 1.68, d. 2,33,
      e. 1.83.
      Pept. for NR, SBR, NBR, BR, IIR.

6 1. Piperidinium-pentamethylenedithiocarbamate
   syn. N-Pentamethyleneammonium-N-pentamethylenedithio-
      carbamate
   2. $(CH_2)_5N.CS.S.H_2N(CH_2)_5$
   3. Accelerator 552   Du Pont
      Accelerator 2P   Anchor
      Nocceler PPD   Ouchi Shinko
      Pentalidine   Prochim
      Robac P.P.D.   Robinson
      Vulkacit P   Bayer
   4. Cream powder. M.W. 246; M.P. 175 °C.; S.G. 1.19.
      Pept. for Neoprene G and KNR types; acc. for NR, NBR, SBR;
      act. for thiuram- and thiazole-type acc.
      △ Acc. 54
      △ Act. 22

7 1. Thioxylenol
   syn. Xylenethiol
      Xylyl mercaptan
   2. $C_6H_3(CH_3)_2SH$

3. Pitt Consol 640   Pitt Consol a
RPA No. 3   Du Pont b
RPA No. 3 conc.   Du Pont c

4. a. 40 % Solution in hydrocarbon solvent. M.W. 138; Flash Point 68.3 °C; S.G. 0.90–0.93.
   b. Amber liquid, (diluent petroleum oil). M.W. 139; Flash Point 74 °C.; S.G. 0.90.
   c. Concentration of RPA No. 3 conc. = 2 × RPA No. 3. Flash-Point 82 °C.; S.G. 1.00.
   Pept. for NR, SBR, IR.

8   1. Zinc 2-benzamidothiophenate

2.

3. Noctizer SZ   Ouchi Shinko
Pepton 65   Cyanamid
Pepton 65   Anchor

4. Off-white powder. M.W. 521.96; M.R. 200–230 °C.; S.G. 1.32.
Pept. for NR, SBR, NBR, IR.

9   1. Zinc benzoyldisulphide
syn. Zinc. thiobenzoate

2. $(C_6H_5COS)_2Zn$

3. Peptizer No. 2   Sanshin a
Robac TBZ   Robinson b

4. Off-white powder.
   a. M.W. 339.75; M.P. 110 °C. min.; S.G. 1.45–1.50.
   b. Hydrate (2 H_2O). M.W. 375.8; M.P. 110 °C.
   Non-staining pept. for NR, SBR, IR.

10   1. Zinc tert butylthiophenate

2.

3. Pitt Consol 651   Pitt Consol a
   Pitt Consol 651 conc.   Pitt Consol b
   Pitt Consol 651 A   Pitt Consol c
4. White powder, extended with inert fillers. M.W. 395; S.G. a. 1.41,
   b. 1.22, c. 1.80.
   Pept. for NR, SBR and other synthetic rubbers.

11   1. Zinc oxide dispersions
2. ZnO
3. Polyzinc   Polychimie
   Grades A, A75, V–01–70
4. Predispersed ZnO.
   Pept. for NR, SBR, NBR, CR and polyblends.

12   1. Zinc salt of pentachlorothiophenol
2. $(C_6Cl_5S-)_2Zn$
3. Ciclizante 1°A   Bozzetto
   Endor   Du Pont
   Renacit IV   Bayer
   Renacit IV/GR   Bayer a
   (Renacit IV 66 %, stearic acid
   and paraffin 34 %)
4. Blueish-gray powder. M.W. 623; M.P. 335 °C. (with decomposition); S.G. 2.33.
   a. Gray granules. M.P. 50–55 °C.; S.G. 1.53.
   Pept. for NR, IIR and SBR (with oil).

13   1. N,S-substituted derivative of o-Aminothiophenol
2. —
3. Peptazin X/BFT   Dimitrova
4. White powder. M.P. 145 °C. min.
   Pept. for NR.

14   1. 2-Naphthalene-thiol / paraffinwax blend (33 : 67)
2. —
3. RPA No. 2   Du Pont
   Premax 2   ACNA Montecatini
4. Cream coloured flakes. M.W. 160.22; Flash Point 160 °C.; S.G.
   0.92.
   Pept. for NR, SBR.

15
1. Sulphonated petroleum products blend
2. —
3. Ancoplas   Anchor
    Grades OB, ER, LP.
4. Pept. for NR, SBR and reclaim rubber.

16
1. o,o'-Dibenzamido-diphenyl disulphide / zinc 2-benzamidothiophenate blend
2. —
3. Noctizer SM   Ouchi Shinko
4. M.P. 130 °C.
    Pept. for NR, IR, BR, SBR.

17
1. Thio-beta-naphthol / inert wax blend (33 : 67)
2. —
3. Vulcamel TBN   ICI †
4. White waxy flakes. M.P. 50 °C.; S.G. 0.9.

18
1. Zinc 2-benzamidothiophenate / zinc thiobenzoate blend
2. —
3. Noctizer SX   Ouchi Shinko
4. M.P. 103 °C.
    Pept. for NR, IR, BR, SBR.

19
1. Composition undisclosed
2. —
3. Aktiplast   Rhein Chemie a
    Aktiplast T   Rhein Chemie a
    Bondogen   Vanderbilt b
    Dispergum N   Grandel c
    Struktol A 50   Schill & Seilacher a
4. a. Yellow powder and pellets. M.P. ca. 80 °C.; S.G. 1.1.
      Pept. for NR and IR, not usable in CR compounds.
   b. Dark brown liquid. S.G. 0.91–0.93.
      General purpose.
   c. Yellow paste. M.R. 70–75 °C.
      Not usable in CR compounds.

Retarders

1   1. Benzoic acid
    2. C_6H_5COOH
    3. Benzoic acid GK   Bayer
       Benzoic acid GV   Bayer
       Retarder BA   Akron
       Ritardante AB   ACNA Montecatini a
    4. White crystalline powder. M.W. 122.12; M.P. 122 °C.; S.G. 1.26.
       Non-staining retarder for NR, SBR, NBR, and their latices.
       a. Benzoic acid with a dispersion agent. M.P. 112–118 °C.;
          S.G. 1.20.

2   1. 4,4'-Diamino diphenyl methane
    2. $CH_2(C_6H_4.NH_2)_2$
    3. Robac 4.4   Robinson a
       Tonox   Uniroyal b
    4. a. Light brown powder. M.W. 198.3; M.R. 75–85 °C.
          Ret. for IIR; acc. for CR; anti-frosting agent for NR.
       b. Brown waxy lump. S.G. 1.18.
       △ Acc. 15
       △ Antidegr. 42

3   1. 2,2'-Dibenzothiazyl disulphide
       syn. Di-(2-benzothiazyl) disulphide
    2.

    3. Accicure MBTS   Alkali
       Altax   Vanderbilt
       Ancatax   Anchor
       Bowax AC/MBTS   Bozzetto
         (MBTS dispersed in Bowax C)
       Eveite DM   ACNA Montecatini
       MBTS   Akron
       MBTS   Cyanamid
       MBTS   Du Pont

MBTS	Naugatuck SpA
MBTS	Uniroyal
Mercasulf MBTS	Bozzetto
Nocceler DM	Ouchi Shinko
Pennac MBTS	Pennwalt
Pneumax DM	Dimitrova
Pneumax F	Dimitrova a
Rapid Accelerator 201	Rhône Poulenc
Sanceler DM	Sanshin
Soxinol DM	Sumitomo
Thiofide MBTS	Monsanto
Vulcafor MBTS	ICI
Vulcafor MBTS	ICI (India)
Vulkacit DM	Bayer

4. Yellowish powder. M.W. 332.50; M.P. 170–175 °C.; S.G. 1.50.
   a. mixture with basic acc.
   Non-staining. Ret. for CR; semi-ultra acc. for NR, NBR, IIR, SBR.
   △ Acc. 92

4 1. Dibutylammoniumoleate

2. $H_3C-CH_2-CH_2-CH_2$
   $H_3C-CH_2-CH_2-CH_2$ $\overset{\oplus}{>}NH_2$ $\overset{\ominus}{O}-\underset{\underset{O}{\|}}{C}-(C_{17}H_{33})$

3. Activator 1102   Anchor
   Barak   Du Pont
   DOB   Organo Synthèse

4. Dark amber liquid. M.W. 409; Flash Point 102 °C.; S.G. 0.88.
   Act.-ret. for thiuram-type acc.; act. for thiazole-, thiuram- and
   sulphenamide-type acc.
   △ Act. 34

5 1. Dicyandiamine / phtalic anhydride blend
   2. —
   3. Retarder AK   Conestoga
   4. White powder. M.P. 123–132 °C.; S.G. 1.48.
      Non-staining ret. for NR, NBR, SBR.

6 1. 2-Mercaptobenzothiazole

2.

3.

Accicure MBT	Alkali
Ancap	Anchor
Captax	Vanderbilt
Eveite M	ACNA Montecatini
MBT	Akron
MBT	Cyanamid
MBT XXX (very pure grade)	Cyanamid
MBT	Du Pont
MBT	Naugatuck SpA
MBT	Uniroyal
Nocceler M	Ouchi Shinko
Pennac MBT	Pennwalt
Pneumax MBT	Dimitrova
Rapid Accelerator 200	Rhône Poulenc
Rotax	Vanderbilt
Sanceler M	Sanshin
Soxinol M	Sumitomo
Thiotax (MBT)	Monsanto
Vulcafor MBT	ICI
Vulcafor MBT	ICI (India)
Vulkacit Merkapto	Bayer

4. Light yellow powder. M.W. 167.25; M.R. 164–175 °C.; S.G. 1.50.
Ret. for CR; non-staining acc. for NR, SBR, NBR, IIR, also for latex; pept. for NR.
△ Acc. 96
△ Pept. 4

7 1. N-Nitroso-diphenylamine

2.

3. Accitard A — Alkali
   Curetard A — Monsanto
   Goodrite Vultrol — Goodrich
   N.D.A. — Prochim
   Redax — Vanderbilt
   Retarder J — Uniroyal
   Retarder 2N — Cyanamid
   Retarder NDPA — Conestoga
   Retrocure — Akron
   Ritardante AN — ACNA Montecatini
   Sconoc — Ouchi Shinko
   Vulcatard A — ICI
   Vulcatard A — ICI (India)
   Vulkalent A — Bayer
   Wiltrol N — Nat. Polychemicals

4. Brown granules. M.W. 198; M.P. 65 °C.; S.G. 1.27.
   Staining ret. for NR, NBR, SBR.

8

1. Phthalic anhydride
2. $C_6H_5(CO)_2O$
3. ESEN — Uniroyal a
   PA — Raschig
   PSA — Raschig
   Retarder AK — Akron
   Retarder B–C — Sanshin a
   Retarder PD — Cyanamid b
   Retarder PD — Anchor b
   Ritardante AF — ACNA Montecatini
   Sconoc 5 (70 % active) — Ouchi Shinko
   Sconoc 7 — Ouchi Shinko
   Sumitard BC — Sumitomo
   Vulkalent B/C — Bayer a
     (surface coated)
   Wiltrol P — Nat. Polychemicals a
4. White powder. M.W. 148.12; M.P. 130 °C. min.; S.G. 1.51.
   a. Surface treated.
   b. Modified.
   Non-staining ret. for NR, SBR, IIR.
   Non-usable in CR-compounds.

9 1. Salicylic acid
   syn. ortho-Hydrobenzoic acid
2. HOC_6H_4OOH
3. Retarder W   Du Pont
   Retarder TSA   Monsanto a
   Ritardante AS   ACNA Montecatini
   Salicylic R   Monsanto
4. Buff powder. M.W. 138.12; M.P. 159 °C.; S.G. 1.37.
   a. White powder. M.P. 155 °C.; S.G. 1.40., with dispersion
   agents.

10 1. Tetrabutyl-thiuram disulphide
2. $[(C_4H_9)_2N.CS.S-]_2$
3. Robac T.B.U.T.   Robinson
   Soxinol TBT   Sumitomo
   TBTS   Bozzetto
4. Brown liquid. M.W. 408.7; solidifies at ca. 20 °C.; S.G. 1.1.
   Insoluble in water.
   Ret. for CR. Non-staining acc. in combination with P.T.D. or
   T.M.T. for NR, SBR, NBR in sulphurless compounds; vulc.
   for NR, SBR, NBR.
   △ Acc. 79
   △ Vulc. 35

Vulcanizing agents

INORGANIC

1 1. Lead oxides

2. —

3.
Mix Lpb 80	Bozzetto a
Mix Pb 80	Bozzetto b
Polyminium	Polychimie b
Polytharge	Polychimie a
grades A, B.D	
RC Granulat PbO	Rhein Chemie a
RC Granulat Pb_3O_4	Rhein Chemie b
—	Anchor c

4. a. PbO dispersion. Acc. for CR; act. for NR, NBR, SBR, IIR; vulc. for CR, CSM.

  b. Pb_3O_4 dispersion. Acc. for CR; act. for IIR.

  c. PbO powder (litharge). Vulc. for CR.

  △ Acc. 1

  △ Act. 1

2 1. Magnesium oxide

2. MgO

3.
Kenmag	Kenrich a
Maglite	Merck b
grades D, L, K, M, Y	
RC Granulat MgO	Rhein Chemie c
(80 % MgO, 20 % saturated	
hydrocarbons and dispersion	
agents)	
Scorchguard	Anchor d
grades C3, O, W	
Scorchguard O	Newalls e
Struktol	Schill & Seilacher f
grades WB 900, WB 902	
(coated)	

4. M.W. 40.32.

  a. S.G. 2.02. Act. for CR.

b. White powders. S.G. 3.3–3.5.
    Vulc. for CR, CSM; act. for SBR and fluoro-elastomers;
    antidegr. for CR, CSM, chlorobutyl, fluoro-elastomers,
    SBR.
c. S.G. 2.06.
d. Grade C3, powder, heavy calcined MgO. Grade O, putty,
    light calcined MgO. Grade W, powder, light calcined MgO.
e. Dispersion. S.G. 2.08.
f. Act. for CR and CSM mixtures.
△ Act. 2
△ Antidegr. 1

3   1. Selenium
    2. Se
    3. Ancasal   Anchor
        Vandex   Vanderbilt
    4. Dark-gray powder. M.P. 217 °C. min.; S.G. 4.80.
        Sec. vulc.

4   1. Sulphur
    2. S
    3. Crystex   Stauffer
        Grades 90, OT (insoluble
        sulphur); tire Brand grades
        21–1, 21–4, 21–7, 21–10; con-
        ditioned tire Brand grades
        21–10TP, 21–12MC, 21–13,
        21–14, 52–AF.
        Insoluble Sulphur 60   Monsanto
        Manox Brand   Anchor
        Grades insoluble sulphur,
        oiled insoluble sulphur, MC
        sulphur (MgCO₃ coated), 9;1
        oil treated sulphur.
        Polysoufre   Polychimie
        Grades SE–01–70, SV–01–80,
        SN–05–75, SA–01–75, IV–01–
        75, IV–01–43.

RC Granulat S	Rhein Chemie

(80% sulphur, 20% saturated hydrocarbons and dispersing agents)

RC Schwefel Extra	Rhein Chemie

(93 % sulphur, 7 % dispersing agents)

S 84	Bozzetto

(predispersed)

Colloidal Sulphur	Bayer

(95 %)

Struktol	Schill & Seilacher

Grades SU 95, (coated, soluble), SU 105, (paste, soluble), SU 106, (coated, insoluble), SU 108, (coated, insoluble), SU 120, (coated, soluble), SU 135, (oil treated, insoluble).

4. —

5 1. Sulphur dichloride
2. SCl_2
3. — Bayer
4. Yellow-to-reddish-yellow liquid. M.W. 102; B.R. 133–141 °C.; S.G. 1.68.
For cold vulcanization of thin-walled rubber goods.

6 1. Tellurium
2. Te
3. Ancatel Anchor
Telloy Vanderbilt
4. Grey powder. M.P. 450 °C.; S.G. 6.26.
Sec. vulc.

7 1. Basic zinccarbonate
2. $ZnCO_3.2 ZnO.3 H_2O$

3. Zinkoxid Transparent   Bayer
   —   Durham
4. White powder. S.G. 3.3–3.5.
Vulc. for CR; act. for sulphur- and peroxide vulcanization of NR, IR, BR, SBR, NBR, IIR, CR, EPDM, CSM.
△ Act. 5

ORGANIC

Phenols

8 1. Alkylphenol disulphide
2. —
3. Vultac   Pennwalt
     Grades 2, 3, 4, 5, 6.
    Vultac   Vondelingenplaat
     Grades 2, 3, 5.
4. Grade 2: dark-brown tacky solid. M.R. 50–60 °C.; S.G. 1.1–1.2.
Grade 3: dark-brown tack-free solid. M.R. 78–93 °C.; S.G. 1.15–1.25.
Grade 4: stearic acid blend. Tacky brown solid. M.R. 48–58 °C.; S.G. 1.05–1.15.
Grade 5: brown powder. S.G. 1.435.
Grade 6: amber-brown liquid. S.G. 1.04–1.09.
Vulc. for NR, SBR, NBR, chlorobutyl- and bromobutyl rubbers.

9 1. Alkylphenol resins
2. —
3. Vulkaresat   Reichhold Albert a
     Grades 510 E, 532 E.
    Vulkaresen   Reichhold Albert b
     Grades E 71, 105 E, 130 E.
    Vulkaresen   American Hoechst
     Grades PA 105, PA 130, PA 510.
4. a. Grade 510 E: M.R. 60–65 °C. Vulc. for EPT.
      Grade 532 E: M.R. 60–65 °C. Vulc. for EPT.

b. Grade E 71: M.R. 60–75 °C. Vulc. for IIR, halogenated IIR, NBR.
Grade 105 E: M.R. 54–60 °C. Vulc. for NR, SBR.
Grade 130 E: M.R. 60–65 °C. Vulc. for NR, SBR, NBR, IIR.

10 1. Bisphenol derivatives
2. —
3. Vulkaresol 315 E   Reichhold Albert
4. 70 % solution in a mixture of butanol and xylene.
Vulc. for all types of rubber.

11 1. Octyl-phenol resin
2. —
3. Varcum 29–530(1198)   Reichhold
4. Light yellow powder. M.R. 80–90 °C.; S.G. 1.03–1.05.
Vulc. for IIR.

12 1. Phenolformaldehyde resin (reactive)
2. —
3. Arrcorez 16   Uniroyal
4. Yellow-to-brown brittle resin in lump form. M.P. 60–70 °C.;
S.G. 1.05.
Vulc. for IIR. FDA appr.

Amines

13 1. Cumenediamine / m-phenylenediamine blend (58 : 42)
2. —
3. Caytur 7   Du Pont
4. Dark-amber-to-black liquid. Freezing Point 16 °C.; S.G. 1.08.
Curing agent for liquid urethane elastomers.

14 1. N,N'-Di-cinnamylidene-1,6-hexanediamine
syn. N,N'-Di-cinnamylidene-1,6-hexamethylenediamine
2.

$-CH=CH-CH=N-(CH_2)_6-N=CH-CH=CH-$

3. Diak No. 3   Du Pont a
Tecnocin A   ACNA Montecatini b

4. Tan coarse powder. M.W. 344; M.R. 82–88 °C.; a. S.G. 1.09,
   b. S.G. 0.374.
   Vulc. for fluoroelastomers.

15 1. 4,4'-Methylene-bis-(2-chloro-aniline)
   syn. 4,4'-Methylene-bis-(o-chloro-aniline)
2. $CH_2(C_6H_3NH_2Cl)_2$
3. Cyanaset M   Cyanamid
   MOCA   Du Pont
4. Light tan pellets. M.W. 267; M.R. 100–109 °C.; S.G. 1.44.
   Curing agent for urethane elastomers.

16 1. Dichlorobenzidine / methylene-bis-(o-chloroaniline) blend
2. —
3. Cyanaset H   Cyanamid a
   Cyanaset S   Cyanamid b
4. Large-grain dark-gray powder. S.G. 1.44; a. M.R. 106–110 °C.;
   b. M.R. 92–100 °C.
   Curing agent for urethane elastomers.

Peroxides

17 1. Benzoyl peroxide
2.
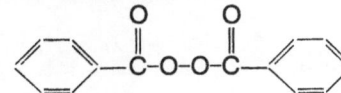
3. Cadox   Nourychem a
   Grades BSD, BSG.
   Lucidol S 50   Noury b
   Norox BZP-S-50   Norac c
4. M.W. 242
   a. Grade BSD: 50 % silicone-oil paste (D.C. 200). S.G. 1.12.
      Grade BSG: 50 % silicone-oil paste (Gen. Electric SF-96–
      1000). S.G. 1.12.
   b. 50 % silicone-oil paste.
   c. 50 % paste in polydimethylsiloxane 1000.
   Vulc. for siliconerubber.

18 1. 1,3-Bis-(tert butyl-peroxy-isopropyl) benzene

2.

$$CH_3-\underset{\underset{CH_3}{|}}{\overset{\overset{CH_3}{|}}{C}}-O-O-\underset{\underset{CH_3}{|}}{\overset{\overset{CH_3}{|}}{C}}-\underset{}{\bigcirc}-\underset{\underset{CH_3}{|}}{\overset{\overset{CH_3}{|}}{C}}-O-O-\underset{\underset{CH_3}{|}}{\overset{\overset{CH_3}{|}}{C}}-CH_3$$

3. Perkadox   Noury a
   Grades 14, 14/40
   Peroximon F 40   ACNA Montecatini b
   Retilox F 40   ACNA Montecatini b

4. a. Grade 14; pale-yellow crystalline powder. M.W. 338; M.P. ca. 50 °C.; S.G. 1.1.
   Grade 14/40, 40 % active ingredient, 60 % $CaCO_3$. Granular. S.G. 1.16.
  b. With inert filler. White small cylinders. M.P. 35 °C.; S.G. 1.43.
Vulc. and crosslinking agent for natural- and synthetic rubbers, siliconerubber and polyolefins.

19 1. Butyl-4,4-bis-(tert butyl peroxy) valerate

2.

$$\begin{array}{c}C(CH_3)_3\\|\\O\\|\\O\qquad\quad O\\|\qquad\quad\|\\H_3C-C-(CH_2)_2-C-O(CH_2)_3CH_3\\|\\O\\|\\O\\|\\C(CH_3)_3\end{array}$$

3. Luperco 230 XL   Lucidol a
   Lupersol 230   Lucidol b

4. a. White powder, 50 % active ingredient on an inert filler. Bulk dens. 24 lbs/cu ft.
  b. Colourless liquid. M.W. 334.46; S.G. 0.9503.
Curing agent for EP rubber, EPT, SBR, silicone rubber and urethane elastomers.

20 1. p-Chlorobenzoyl peroxide

2.

$$Cl-\langle\rangle-\overset{O}{\overset{\parallel}{C}}-O-O-\overset{O}{\overset{\parallel}{C}}-\langle\rangle-Cl$$

3. Cadox PS   Nourychem

4. 50 % Silicone-oil paste, containing 10 % dibutylphthalate. S.G. 1.17.
   Curing agent for silicone rubber.

21 1. Cumene hydroperoxide

2.

$$\langle\rangle-\overset{CH_3}{\underset{CH_3}{\overset{\mid}{\underset{\mid}{C}}}}-O-OH$$

3. Trigonox K 70   Noury

4. Colourless liquid, 70% active ingredient in a mixture of alcohols, ketones and cumene. Flash Point 61 °C.; S.G. 1.01–1.04.
   Curing agent for polysulphide rubber.

22 1. tert Butyl cumyl peroxide

2.

$$CH_3-\overset{CH_3}{\underset{CH_3}{\overset{\mid}{\underset{\mid}{C}}}}-O-O-\overset{CH_3}{\underset{CH_3}{\overset{\mid}{\underset{\mid}{C}}}}-\langle\rangle$$

3. Trigonox   Noury
   Grades T, TV–50

4. Grade T: Colourless liquid. M.W. 208, M.P. <17 °C.; S.G. 0.96.
   Grade TV–50: White powder, 50 % active ingredient with silica filler. S.G. 1.2.
   Vulc. for NR, synthetic rubbers, silicone rubber and polyolefins.

23 1. tert Butyl perbenzoate

2.

$$CH_3-\overset{CH_3}{\underset{CH_3}{\overset{\mid}{\underset{\mid}{C}}}}-O-O-\overset{O}{\overset{\parallel}{C}}-\langle\rangle$$

3. Trigonox C   Noury

4. Clear slightly yellow liquid. M.W. 194; Flash Point 50 °C.; S.G. 1.04.
   Vulc. for silicone rubber.

24 1. Di-tert butyl peroxide
2.

$$CH_3-\underset{\underset{CH_3}{|}}{\overset{\overset{CH_3}{|}}{C}}-O-O-\underset{\underset{CH_3}{|}}{\overset{\overset{CH_3}{|}}{C}}-CH_3$$

3. —   Wallace & Tiernan a
   Trigonox B   Noury
4. Colourless, clear liquid. M.W. 146; B.P. 111 °C.; Flash Point 8 °C.; S.G. 0.793.
   a. B.P. 119 °C.
   Vulc. for natural- and synthetic rubber, silicone rubber and polyolefins.

25 1. 3,3-Di-tert butyl-peroxy-butanecarboxylic-n-butyl-ester
2.

$$H_3C-\underset{\underset{O}{|}}{\overset{\overset{CH_3}{|}}{C}}-CH_3$$
$$\underset{O}{|}$$
$$H_3C-\underset{\underset{O}{|}}{\overset{\overset{O}{|}}{C}}-CH_2-CH_2-\overset{\overset{O}{||}}{C}-O-C_4H_9$$
$$\underset{O}{|}$$
$$H_3C-\underset{\underset{CH_3}{|}}{\overset{\overset{CH_3}{|}}{C}}-CH_3$$

3. Trigonox   Noury
   Grades 17, 17/40.
4. Grade 17: Pale yellow liquid. M.W. 334; S.G. 0.95.
   Grade 17/40: White powder, 40% active ingredient with $CaCO_3$. S.G. 1.56.
   Vulc. for natural- and synthetic rubber, silicone rubber and polyolefins.

26 1. 1,1-Di-tert butyl-peroxy-3,3,5-trimethyl cyclohexane

2.

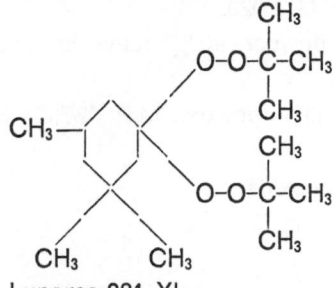

3. Luperco 231–XL   Wallace & Tiernan
   (40 % on inert filler)
   Luperco 231–XLP   Wallace & Tiernan
   (30 % paste)
   Luperox 231   Wallace & Tiernan
   (93 % liquid)
   Percadox 29/40   Noury
   (40 % on $CaCO_3$)
   Trigonox 29/40   Noury

4. M.W. 302.44.
   Vulc. for natural- and synthetic rubber, and for silicone rubber.

27 1. Dicumyl peroxide

2.

3. —   Wallace & Tiernan a
   Di-Cup   Hercules b
      Grades 40C, 40KE, R, T
   Perkadox   Noury c
      Grades BC 40, SB

4. M.W. 270; M.P. 42 °C.
   a. 95 % active ingredient on inert filler; 40 % active ingredient
      on inert filler.
   b. Grade 40C: White powder, 39–41 % active ingredient. S.G.
      1.53.
      Grade 40KE: White powder, 39.5–41.5 % active ingredient.
      S.G. 1.607.
      Grade R: White granules, 96–100 % active ingredient. M.P.
      38 °C.; S.G. 1.018.

Grade T: White crystalline powder, 90–93 % active ingredient. M.P. 30 °C.; S.G. 1.023.

c. Grade BC 40: White powder, 40 % active ingredient. S.G. 1.53.

Grade SB: White crytalline powder. M.P. 42 °C.; S.G. 1.02.

28 1. Di-(2,4-dichlorobenzoyl) peroxide

2.

3. Cadox TS 50   Nourychem a
   Luperco   Wallace & Tiernan b
     Grades CDB, CST.
   Norox DBP–S–50   Norac c
   Perkadox   Noury d
     Grades PDB 50, PDS 50, SD.
   Siloprene Crosslinking Agent   Bayer e
     CL 40

4. M.W. 380.0.
   a. 50 % Paste in dibutylphthalate and silicone oil. S.G. 1.24.
   b. Grade CDB: 50 % paste in dibutylphthalate.
      Grade CST: 50 % paste in silicone oil.
   c. 50 % Paste in polydimethylsiloxane 1000.
   d. Grade PDB 50: 50 % paste in dibutylphthalate.
      Grade SD: white powder, 95 % active ingredient.
   e. 40 % Paste in silicone oil. S.G. 1.18.
   Vulc. for silicone rubber.

29 1. 2,5-Dimethyl-2,5-(di-tert butyl-peroxy) hexane

2.

3. Luperco 101 XL   Lucidol a
   Luperco 101 XL   Pennwalt a
   Luperco 101 XL   Wallace & Tiernan a
   Lupersol 101   Lucidol b
   Lupersol 101   Pennwalt b
   Lupersol 101   Wallace & Tiernan b
   Varox   Vanderbilt c
4. M.W. 290.45.
   a. White powder, 45–50 % active ingredient on inert carrier.
      S.G. ca. 1.50.
   b. Water-white liquid. B.P. 250 °C.; S.G. 0.87.
   c. Liquid and powder.
   Crosslinking agents for SBR, EPDM, PU, and silicone rubber.

30 1. 2,5-Dimethyl-2,5-(tert butyl-peroxy) hexyne-3
        2.

$$
\begin{array}{ccc}
CH_3 & & CH_3 \\
| & & | \\
CH_3-C-C\equiv\equiv\equiv C-C-CH_3 \\
| & & | \\
O & & O \\
| & & | \\
O & & O \\
| & & | \\
CH_3-C-CH_3 & & CH_3-C-CH_3 \\
| & & | \\
CH_3 & & CH_3
\end{array}
$$

3. Luperco 130 XL   Wallace & Tiernan a
   Luperco 130 XL   Lucidol a
   Luperox 130   Wallace & Tiernan b
   Lupersol 130   Lucidol b
4. a. White powder, 45 % active ingredient. Bulk density 0.481.
   b. Light-yellow liquid. M.W. 286.42; Flash Point 95 °C.; B.P.
      243 °C.
   Crosslinking agents for EPR.

31 1. 4-Methyl-2,2-bis-(tert butyl-peroxy) pentane
        2. $CH_3-CH-CH_2-C-CH_3$

$$
\begin{array}{cc}
| & \diagup\diagdown \\
CH_3 & O \quad O \\
& | \quad | \\
& O \quad O \\
& | \quad | \\
(CH_3)_3C & C(CH_3)_3
\end{array}
$$

3. Luperco 144 XL   Wallace & Tiernan
4. Powder, 40 % active ingredient on inert carrier.
   Vulc. for EPR, EPDM.

Thiuram-sulphides

---

32 1. Dipentamethylene-thiuram disulphide
2. $(C_5H_{10}N.CS.S)_2$
3. Robac P.T.D.   Robinson a
   Robac P.T.D. 86   Robinson b
   (containing extra sulphur)
4. Cream coloured powder. M.W. 320.6.
   a. M.P. 120 °C.
   b. M.P. 110 °C.
   Non-staining vulc. and acc. for latex (gloves) and for IIR
   (pharmaceutical closures).
   △ Acc. 75

33 1. Dipentamethylene-thiuram hexasulphide
2.

$$\underset{\text{CH}_2\text{-CH}_2}{\overset{\text{CH}_2\text{-CH}_2}{\diagup\diagdown}} \text{H}_2\text{C} \diagdown\diagup \underset{\text{CH}_2\text{-CH}_2}{\overset{\text{S}\quad\quad\text{S}}{\text{N-C-(S)}_6\text{-C-N}}} \diagup\diagdown \underset{\text{CH}_2\text{-CH}_2}{\overset{\text{CH}_2\text{-CH}_2}{\text{CH}_2}}$$

3. DPTT   Akron
   Sulfads   Vanderbilt
   Tetrone A   Du Pont
4. Light gray powder. M.W. 448; M.P. 110 °C.; S.G. 1.53.
   Vulc. for IR; non-staining acc. for NR, SBR, NBR, CR, IR, IIR,
   EPDM, CSM. Also for latex.
   △ Acc. 77

34 1. Dipentamethylene-thiuram-tetrasulphide
2. $(CH_2)_5N.CS.S_4.CS.N(CH_2)_5$
3. Accelerator 4P   Anchor
   DPTT   Hasselt
   Nocceler TRA   Ouchi Shinko
   Robac P.25   Robinson
   Soxinol TRA   Sumitomo

4. Light yellow powder or rods. M.W. 384.69; M.P. 115 °C.; S.G. 1.50.
   Vulc. and ultra-acc. for CSM, IIR, EPDM, NBR, SBR, IR, CR, NR; act. for thiazole- and sulphenamide-type acc.
   △ Acc. 76
   △ Act. 7

35 1. Tetrabutyl-thiuram disulphide
   2. $[(C_4H_9)_2N.CS.S-]_2$
   3. Robac T.B.U.T.   Robinson
      Soxinol TBT   Sumitomo
      TBTS   Bozzetto
   4. Brown liquid. M.W. 408.7; solidifies at ca. 20 °C.; S.G. 1.1. Insoluble in water.
      Vulc. for NR, SBR, NBR; non-staining acc. in combination with P.T.D. or T.M.T. for NR, SBR, NBR in sulphurless compounds; ret. for CR.
      △ Acc. 79
      △ Ret. 10

36 1. Tetraethyl-thiuram-disulphide
   2. $(C_2H_5)_2N.CS.S.S.CS.N(C_2H_5)_2$
   3. Accicure TET   Alkali
      Aceto TETD   Aceto
      Ancazide ET   Anchor
      Ethyl Thiram   Pennwalt
      Ethyl Thiurad   Monsanto
      Ethyl Tuads   Vanderbilt a
      Ethyl Tuex   Naugatuck SpA
      Ethyl Tuex   Uniroyal
      Etiurac   Ticino
      Eveite T   ACNA Montecatini
      Hermat TET   Dimitrova
      Nocceler TET   Ouchi Shinko
      Robac TET   Robinson
      Sanceler TET   Sanshin
      Soxinol TET   Sumitomo
      Superaccelerator 481   Rhône Poulenc

TETD	Hasselt
TETD	Prochim
TETS	Bozzetto
Thiuram E	Du Pont
Vondac TET	Vondelingenplaat
Vulcafor TET	ICI
Vulcafor TET	ICI (India)

4. Grayish-white powders and pellets. M.W. 296.54; M.P. 71–73 °C.;
   S.G. 1.26.
   a. Powder and rods.
   Vulc. for sulphurless compounds; acc. for NR, SBR, NBR, IIR,
   IR, EPDM; act. for thiazole-, guanidine- and aldehyde-type
   acc.; stabilizer for Neoprene GN. Non-staining.
   △ Acc. 81
   △ Act. 8
   △ Antidegr. 96

37 1. Tetramethyl-thiuram disulphide
   2. $(CH_3)_2N.CS.S.S.CS.N(CH_3)_2$
   3.

Accicure TMT	Alkali
Aceto TMTD	Aceto
Ancazide ME	Anchor
Cyuram DS	Cyanamid
Eveite 4MT	ACNA Montecatini
Hermat TMT	Dimitrova
Methyl Thiram (oiled and extruded)	Pennwalt
Methyl Tuads	Vanderbilt a
Metiurac	Ticino
Nocceler TT	Ouchi Shinko
RC Granulat TMTD (80 % TMTD, 20 % saturated hydrocarbons and dispersing agents)	Rhein Chemie
Robac TMT	Robinson
Sanceler TT	Sanshin
Soxinol TT	Sumitomo
Superaccelerator 501	Rhône Poulenc

Thiurad	Monsanto
Thiuram 16	Metallgesellschaft
Thiuram M	Du Pont
TMTD	Hasselt
TMTD	Akron
TMTD	Prochim
TMTS	Bozzetto
TMTS oleato	Bozzetto
(83 % TMTS, 17 % paraffin oil)	
Tuex	Naugatuck SpA
Tuex	Uniroyal
Vondac TMT	Vondelingenplaat
Vulcafor TMT	ICI
Vulcafor TMT	ICI (India)
Vulkacit Thiuram	Bayer
Vulkacit Thiuram C	Bayer
(surface coated)	
Vulkacit Thiuram GR	Bayer
(granules)	

4. White to yellow powder. M.W. 240.44; M.P. 146–148 °C.; S.G. 1.3–1.4.

a. Powder and rods.

Non-staining ultra-acc. and vulc.; act. for thiazole- and sulphenamide-type acc.

△ Acc. 83

△ Act. 9

38 1. Tetraethyl-thiuram disulphide / tetramethyl-thiuram disulphide blend

2. —

3. Methyl Ethyl Tuads   Vanderbilt a
   Pennac TM   Pennwalt b

4. a. 50 : 50 mixture. White to-cream rods. M.P. 62 °C. min.; S.G. 1.32.

b. Light buff powder. M.W. 257.5; S.G. 1.24.

Vulc. and acc. for sulphurless or low-sulphur stocks of NR, SBR, EPDM.

△ Acc. 84

Carbamates

39 1. Hexamethylene-diaminecarbamate

2.

$$H_3\overset{\oplus}{N}-(CH_2)_6-\underset{H}{N}\diagup\overset{\overset{\ominus}{C}O_2}{}$$

3. Diak No. 1   Du Pont
   Tecnocin B   ACNA Montecatini
4. White powder. M.W. 160; M.R. 152–155 °C.; S.G. 1.15.
   Vulc. for fluoroelastomers.

40 1. Selenium diethyldithiocarbamate

2. $[(C_2H_5)_2N.CS.S-]_4Se$

3. Ethyl Selenac   Vanderbilt
   Ethyl Seleram SA–66–1   Pennwalt
      (oiled)
   Seleniame   Prochim
   Soxinol SE   Sumitomo
4. Yellow-orange powder. M.W. 672; M.R. 59–85 °C.; S.G. 1.32.
   Vulc. for NR, NBR, SBR; acc. for IIR; act. for thiazole-type acc.
   △ Acc. 57
   △ Act. 23

41 1. Selenium dimethyldithiocarbamate

2. $[(CH_3)_2N.CS.S-]_4Se$

3. Methyl Selenac   Vanderbilt
4. Yellow powder and rods. M.W. 559.78; M.R. 140–172 °C.; S.G.
   1.58.
   Vulc. and acc. for NR, IIR, SBR, BR, IR.
   △ Acc. 58

Miscellaneous, mixtures and undisclosed compositions

42 1. Dibenzoyl-p-quinonedioxime

2. $C_6H_5-CO-O-N=C_6H_4=N-O-CO-C_6H_5$

3. Dibenzo G–M–F   Uniroyal
   Dibenzo PQD   Conestoga
   Kencure Dibenzo Q.D.O.   Kenrich
   Vulnoc DGM   Ouchi Shinko

Vulnoc DGM paste   Ouchi Shinko
(50 % active ingredient)
4. Gray-brown powder. M.W. 346; S.G. 1.37; decomposition above 200 °C.
Vulc. for NR, SBR, IR, IIR.

43 1. dimeric 2,4-Toluylenediisocyanate
    2.

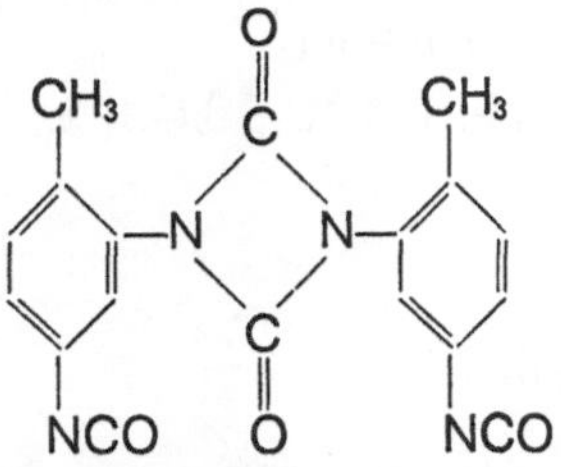

3. Desmodur TT   Bayer
4. White powder. M.W. 348; M.P. 145 °C. min.; S.G. 1.48.
Curing agent for urethane elastomers.

44 1. 4,4'-Dithiodimorpholine
syn. Dimorpholinyl disulphide
      morpholine disulphide
    2.

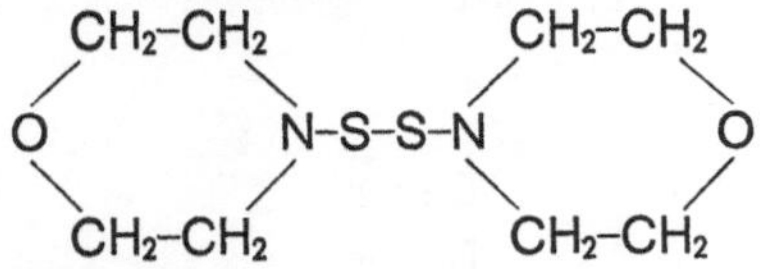

3. Deovulc M   Grandel
    Sulfasan R   Monsanto
    Vanax A   Vanderbilt
    Vondac DTDM   Vondelingenplaat
    Vulnoc R   Ouchi Shinko
4. White to grey powder. M.W. 236.36; M.R. 115–127 °C.; S.G. 1.35.
FDA appr.
Vulc. and acc. for NR, SBR, NBR, IIR, EPDM.
△ Acc. 85

45 1. N,N'-Di-thio-bis-(hexahydro-2H-azepinon-2)
    2.

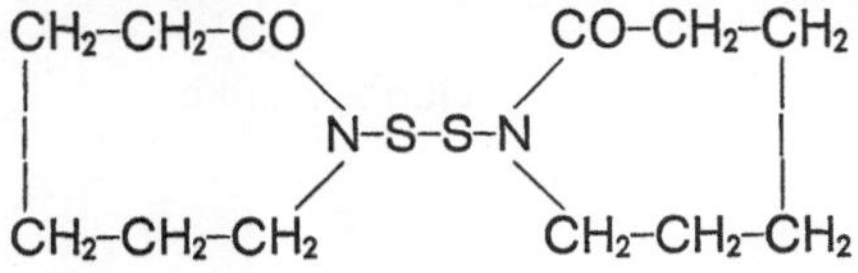

3. Rhenocure S   Rhein Chemie
4. Cream powder or granules. M.W. 288; S.G. 1.3.
   Non-staining vulc. for natural and synthetic rubber.

46
1. Hydroquinone (treated)
2. $C_6H_4(OH)_2$
3. Diak No. 5   Du Pont
4. Off-white powder. M.W. 110; M.P. 170 °C.; S.G. 1.3.
   Curing agent for Viton D-80.

47
1. Hydroquinone-dihydroxyethylether
2.

$$HO-CH_2-CH_2-O-\!\!\left\langle\!\!\bigcirc\!\!\right\rangle\!\!-O-CH_2-CH_2-OH$$

3. Crosslinking Agent 30/10   Bayer
4. White crystalline powder. M.W. 198; M.P. 104 °C.; S.G. 1.34.
   Sec. crosslinking agent for urethane elastomers in combination
   with Desmodur TT (dimeric 2,4-toluylenediisocyanate).

48
1. organic Lead salts
2. —
3. Desmorapid 600   Bayer
   Desmorapid DA   Bayer
4. Yellow powder. S.G. 2.
   Crosslinking agents for urethane elastomers.

49
1. Poly-p-dinitrosobenzene
2.

$$\left[\begin{array}{c} N=O \\ \bigcirc \\ N=O \end{array}\right]_n$$

3. Vulnoc DNB   Ouchi Shinko
4. M.W. $(136.11)_n$
   Vulc. for IIR.

50 1. Poly-2,2,4-trimethyl-1,2-dihydroquinoline

2.
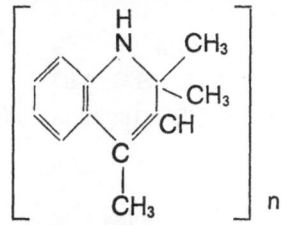

3.

Aceto POD	Aceto
Agerite AK	Anchor
(tradename outside U.K.	
Antioxidant 184)	
Agerite Resin D	Vanderbilt
Anox HB	Bozzetto
Antigene RD	Sumitomo
Antioxidant HS	Bayer a
Flectol H	Monsanto
Nocrac 224	Ouchi Shinko
Pennox HR	Pennwalt
Permanax 45	Rhône Poulenc

4. Buff powder or flakes. M.P. 100–120 °C. (a. M.P. 75 °C. min.); S.G. 1.08.
   Vulc. for CR; semi-staining oxi, heat, inh for NR, NBR, BR, IR, CR. Also for latex.
   △ Antidegr. 95

51 1. p-Quinone-dioxime

2.
$$HO-N=\langle\!\!=\!\!=\rangle=N-OH$$

3.

Curing Agent CDO 50	Bayer
G–M–F	Uniroyal
Kencure QDO	Kenrich
PQD	Conestoga
Vulcafor BQ	ICI †
Vulnoc GM	Ouchi Shinko
Vulnoc GM paste	Ouchi Shinko
(50 % active ingredient)	

4. Dark-brown powder. M.W. 138.13; M.P. 215 °C.; S.G. 1.97.
   Vulc. for NR, SBR, IIR and polysulphiderubber.

52 1. Tetrachloro-p-benzoquinone
2. $O:C_6Cl_4:O-1,4$
3. Vulklor   Uniroyal
4. Yellow powder. M.W. 244; M.P. 290 °C.; S.G. 1.97.
Non-staining vulc. for NR, NBR, IIR, SBR. (sulphurless vulcanization)

53 1. Zincchloride - benzothiazyl disulphide complex
2. —
3. Caytur 4   Du Pont
(formerly LD–755)
4. Yellow powder. M.P. 235 °C.; S.G. 1.85.
Crosslinking agent for sulphur-curable urethane-elastomers.

54 1. Composition undisclosed
2. —
3. Diak No. 4   Du Pont a
Kenmix   Kenrich b
Synform C 1000   Anchor c
Trigonox X–17/40   Noury d
4. a. White powder. M.R. 145–155 °C.; S.G. 1.23.
Curing agent for fluoro-elestomers.
b. △ Acc. 141
△ Act. 44
△ Antidegr. 125
c. Orange-yellow lumps. M.P. 65 °C.
Curing agent for IIR.
d. Powder.
General purpose.

Subject index

N,N'-Di-thio-bis-(hexahydro-2H-azepinon-2). Vulc. 45
Dithiocarbamates (activated). Acc. 40
Dithiocarbamate / tetraethyl-thiuram disulphide blend. Acc. 117
4,4'-Dithiodimorpholine. Acc. 85, Vulc. 44
N,N'-Di-o-tolylguanidine. Acc. 32, Act. 38
Di-o-tolylguanidine salt of di-catecholborate. Acc. 33, Antidegr. 116
N,N'-Di-o-tolyl-thiourea. Acc. 26
6-Dodecyl-1,2-dihydro-2,2,4-trimethyl-quinoline. Antidegr. 91

6-Ethoxy-2,2,4-trimethyl-1,2-dihydroquinoline. Antidegr. 92
Ethylchloride - formaldehyde - ammonia reaction product. Acc. 3
Ethylenedimethacrylate. Ca. 2
Ethyleneglycoldimethacrylate. Ca. 2
Ethylene-thiourea. Acc. 27, Antidegr. 93

Ferric dimethyldithiocarbamate. Acc. 51

Glycol-dimercaptoacetate. Acc. 118

Heptaldehyde - aniline condensation product. Acc. 17, Act. 14
Hexamethylene-diaminecarbamate. Vulc. 39
Hexamethylenetetramine (with additives). Acc. 2, Act. 15
1,6-Hexanediol-bis-[3-(3',5'-di-*tert* butyl-4-hydroxyphenyl) propionate].
    Antidegr. 20
Hydroquinone (treated). Vulc. 46
Hydroquinone-dihydroxyethylether. Vulc. 47
Hydroquinone-dimethylether. Antidegr. 112
Hydroquinone-monobenzylether. Antidegr. 117
Hydroquinone-monomethylether. Antidegr. 118
o-Hydroxybenzoic acid. Ret. 9

p-Isopropoxy-diphenylamine. Antidegr. 70
4-Isopropylamino-diphenylamine. Antidegr. 57
N-Isopropyl-N'-phenyl-p-phenylenediamine. Antidegr. 57

Lead dimethyldithiocarbamate. Acc. 52, Act. 21
Lead oxides. Acc. 1, Act. 1, Vulc. 1
Lead pentamethylenedithiocarbamate. Acc. 53
Organic Lead salts. Vulc. 48

Trade names index

Accinox SP	Antidegr. 37
Accitard A	Ret. 7
Aceto AN	Antidegr. 78
Aceto DIPP	Antidegr. 54
Aceto HMT	Acc. 2, Act. 15
Aceto Ozone	Antidegr. 46
Aceto PAN	Antidegr. 79
Aceto PBN	Antidegr. 82
Aceto POD	Antidegr. 95, Vulc. 50
Aceto TETD	Acc. 81, Act. 8, Antidegr. 96, Vulc. 36
Aceto TMTD	Acc. 83, Act. 9, Vulc. 37
Aceto TMTM	Acc. 82, Act. 10
Aceto ZDBD	Acc. 65, Act. 28, Antidegr. 103
Aceto ZDED	Acc. 67, Act. 29
Aceto ZDMD	Acc. 68, Act. 30
AC Polyethylene 6	Antidegr. 2
AC Polyethylene 6A	Antidegr. 2
AC Polyethylene 7	Antidegr. 2
AC Polyethylene 8	Antidegr. 2
AC Polyethylene 8A	Antidegr. 2
AC Polyethylene 615	Antidegr. 2
AC Polyethylene 617	Antidegr. 2
Activator 1102	Act. 34, Ret. 4
Activator 1203	Act. 42
Activator DN	Act. 42
Activox	Acc. 6
Additiv-Paste 1100	Act. 44
Additiv-Paste 1600	Act. 44
Aero Xanthate 343	Acc. 36
Agerite AK	Antidegr. 95, Vulc. 50
Agerite Alba	Antidegr. 117
Agerite DPPD	Antidegr. 56
Agerite GEL	Antidegr. 77
Agerite Geltrol	Antidegr. 22
Agerite Hipar	Antidegr. 86
Agerite HP	Antidegr. 84
Agerite HPX	Antidegr. 84
Agerite ISO	Antidegr. 70

Eveite 303	Acc. 95
Eveite 404	Acc. 127
Eveite 1000	Acc. 34, Act. 40
Eveite A	Acc. 16, Antidegr. 43
Eveite Butil Z	Acc. 65, Act. 28, Antidegr. 103
Eveite D	Acc. 31, Act. 36
Eveite DM	Acc. 92, Ret. 3
Eveite DOT	Acc. 26
Eveite DOTG	Acc. 32, Act. 38
Eveite F	Acc. 108
Eveite K	Acc. 61
Eveite L	Acc. 60, Act. 25
Eveite M	Acc. 96, Pept. 4, Ret. 6
Eveite Metil Z	Acc. 68, Act. 30
Eveite MS	Acc. 103
Eveite MST	Acc. 82, Act. 10
Eveite MZ	Acc. 100, Antidegr. 99
Eveite P	Acc. 70
Eveite T	Acc. 81, Act. 8, Antidegr. 96, Vulc. 36
Eveite TC	Acc. 25, Act. 37
Eveite UR	Acc. 2, Act. 15
Eveite XZ	Acc. 38
Eveite Z	Acc. 67, Act. 29
Ficel (Genitron)	Ba. 2, Ba. 3, Ba. 8, Ba. 9, Ba. 13, Ba. 16
Flectol H	Antidegr. 95, Vulc. 50
Flexamine	Antidegr. 107
Flexamine G	Antidegr. 107
Flexzone 3C	Antidegr. 57
Flexzone 4L	Antidegr. 49
Flexzone 5L	Antidegr. 46
Flexzone 6H	Antidegr. 59
Flexzone 7L	Antidegr. 53
Flexzone 8L	Antidegr. 47
Flocculent	Act. 6
Genitron AC	Ba. 2
Genitron AZDN	Ba. 3

Polytharge B	Acc. 1, Act. 1, Vulc. 1
Polytharge D	Acc. 1, Act. 1, Vulc. 1
Polyzinc A	Pept. 11
Polyzinc A 75	Pept. 11
Polyzinc V-01-70	Pept. 11
Porofor ADC/R	Ba. 2
Porofor B 13/CP 50	Ba. 7
Porofor-BSH-Paste	Ba. 8
Porofor-BSH-Paste M	Ba. 8
Porofor-BSH-Powder	Ba. 8
Porofor DNO/F	Ba. 6
Porofor DNO/N	Ba. 6
Porofor N	Ba. 3
PQD	Vulc. 51
Premax 2	Pept. 14
PSA	Ret. 8
R-2 Crystals	Acc. 87
Raluquin K	Antidegr. 125
Rapid Accelerator 200	Acc. 96, Pept. 4, Ret. 6
Rapid Accelerator 201	Acc. 92, Ret. 3
Rapid Accelerator 205	Acc. 100, Antidegr. 99
Rapid Accelerator 300A	Acc. 14, Act. 11, Antidegr. 41
Rapid Accelerator 465	Acc. 141
RC Granulat DETU	Acc. 23, Antidegr. 110
RC Granulat HEXA	Acc. 2, Act. 15
RC Granulat IPPD	Antidegr. 57
RC Granulat MgO	Act. 2, Antidegr. 1, Vulc. 2
RC Granulat PbO	Acc. 1, Act. 1, Vulc. 1
RC Granulat Pb_3O_4	Acc. 1, Act. 1, Vulc. 1
RC Granulat S	Vulc. 4
RC Granulat TMTD	Acc. 83, Act. 9, Vulc. 37
RC Granulat ZnO	Act. 6
RC Lichtschutz- mittel 520	Antidegr. 2
RC-Schwefel Extra	Vulc. 4
RC-Zinkoxid 64	Act. 6
Redax	Ret. 7

Vulkaresol 315 E	Vulc. 10
Vulklor	Vulc. 52
Vulnoc DGM	Vulc. 42
Vulnoc DGM Paste	Vulc. 42
Vulnoc DNB	Vulc. 49
Vulnoc GM	Vulc. 51
Vulnoc GM Paste	Vulc. 51
Vulnoc R	Acc. 85, Vulc. 44
Vultac 2	Vulc. 8
Vultac 3	Vulc. 8
Vultac 4	Vulc. 8
Vultac 5	Vulc. 8
Vultac 6	Vulc. 8
Wiltrol N	Ret. 7
Wiltrol P	Ret. 8
Wingstay 100	Antidegr. 50
Wingstay 200	Antidegr. 50
Wingstay 250	Antidegr. 63
Wingstay 275	Antidegr. 63
Wingstay 300	Antidegr. 58
Wingstay L	Antidegr. 22
Wingstay S	Antidegr. 37
Wingstay T	Antidegr. 5
Wingstay V	Antidegr. 6
Wytox 345	Antidegr. 125
Wytox ADP	Antidegr. 69
Wytox ADP-X	Antidegr. 69
Zalba Special	Antidegr. 22
ZBEC	Acc. 64, Act. 27
ZDBC	Acc. 65, Act. 28, Antidegr. 103
ZDEC	Acc. 67, Act. 29
ZDMC	Acc. 68, Act. 30
Zenite	Acc. 100, Antidegr. 99
Zenite A	Acc. 133
Zenite AM	Acc. 133
Zenite Special	Acc. 100, Antidegr. 99

Contents

NOTES

NOTES